AF883101

CHRISTIAN LASSEN

How to Change the World: One Person at a Time

Rising from Homo Sapiens to Human Beings

Westbury Publishing

First published by Westbury Publishing 2019

Copyright © 2019 by Christian Lassen

All rights reserved. No part of this publication may be reproduced, stored or transmitted in any form or by any means, electronic, mechanical, photocopying, recording, scanning, or otherwise without written permission from the publisher. It is illegal to copy this book, post it to a website, or distribute it by any other means without permission.

Christian Lassen asserts the moral right to be identified as the author of this work.

Christian Lassen has no responsibility for the persistence or accuracy of URLs for external or third-party Internet Websites referred to in this publication and does not guarantee that any content on such Websites is, or will remain, accurate or appropriate.

Designations used by companies to distinguish their products are often claimed as trademarks. All brand names and product names used in this book and on its cover are trade names, service marks, trademarks and registered trademarks of their respective owners. The publishers and the book are not associated with any product or vendor mentioned in this book. None of the companies referenced within the book have endorsed the book.

While this book aims to help as many people as possible, it is a book of general ideas and should not be construed to be catered to your individual circumstances. It is not a replacement for the specialized help of trained professionals in your life. It is not meant to be a cure, treatment, or therapy in any professional way for anyone's individual ailments, problems, or illnesses. Christian Lassen, PLLC cannot accept responsibility for the way a reader may or may not appropriately apply the ideas in this book to their own individual circumstances.

First edition

ISBN: 978-1-7338511-9-0

This book was professionally typeset on Reedsy.
Find out more at reedsy.com

To my wife Bronwyn, first, for being the reason to hope for the future.

To my children for being the reasons to stick out the hard times.

To all the amazingly smart people who I've learned from.

And lastly, to everyone who's ever struggled and shared their stories.

In a sudden embarrassment for words I said, "Do you collect shells?"

"Only ones like this," he said softly, gesturing amidst the wreckage of the shore, "and only for the living." He stooped again, and skipped another star neatly across the water. "The stars," he said, "throw well. One can help them."

Loren Eisely
The Star Thrower

Contents

II How to Make Change Happen. How to Love People, Including Yourself

Preface

Parent's Guide

This book is meant for people who can understand "grown up" things. There is nothing explicit, vulgar, graphic, or detailed in it. Vocabulary and detail-wise it is very "PG." However, the topics discussed are meant for audiences who can comprehend the reality of human nature and all of its behaviors. While every parent, child, and family relationship is different, I'd probably hold off on my kids reading this until they are around 10-13 years old, though, in reality, they probably wouldn't care about reading it that young anyway. Whether they read it or not, that's about the age I would start discussing these types of things with them to prepare them for what lies ahead.

Acknowledgement

To everyone who's ever come before me, who've figured out how to do things the hard way and lived to tell others about it. To those who've learned to watch the world around them and notice the subtle patterns that drive the universe. To those who didn't quit or shrink, while facing death and despair in the face. To all those who have kept Hope alive and cleared the way for those behind them.

I

Understanding Who You Are and Why You Do What You Do

Humanity is stuck in an awkward position. Many of us recognize our own animal nature while holding the rest of the world to much higher standards. We are all caught between the natural and the supernatural, between Biology and Ideology. We are torn in two. This first section focuses on our natural instincts and drives. The biology and instinct. What our nature tells us. When we know where we've been, it helps us plan for the future.

Feel free to read this section now, or after you read Part 2.

1

Introduction or ... Why Does Change NEED to Happen

I'd like to have you imagine something as you begin this book. Read the next paragraph slowly, and take the time to imagine the scene in your head.

Imagine laying in bed, the day after a massive test or presentation. You spent all night tossing and turning worrying about how well you did. Your neck feels stiff and sore. The night seems to be endless as your brain just keeps replaying the same thoughts, over and over and over. After what seems like an eternity, the sun begins to rise, lighting your bedroom through the window and you can't believe you were awake all night without getting any sleep. Some music begins to play from nearby. It's not your favorite music, but it's still the kind of music that makes you smile every time you hear it. As you smile, you feel your body soften a bit. Your head sinks into your pillow just a bit further. Your muscles go limp. Your eyes gently close. You begin to feel your bed envelop you like the soft pillowy pad that it is. You drift off to sleep, finally.

If you are like most people, the first few sentences of that paragraph might have made you tense up, ever so slightly. Your heart might have quickened its pace, just a few beats. As you kept reading and imagined relaxing, there's a good chance you really did relax, and your heart rate really did slow down. I had to take the risk of putting you to sleep right at the beginning of this book, but I wanted you to notice something.

Why did some words, written by a stranger, on a page in a book, have an effect on your body? How was I able to reach across time and space to you and have an influence on your heart rate and muscle tension? How much control does our environment have on each of us? I'm going to tell you that it's a lot. More than you will want to admit. You have very little REAL control over your body, though that little tiny bit is what matters. It matters more than anything else in this universe.

By helping you recognize where you can take control of yourself, I can change the world, one person at a time. And by you helping yourself, you have changed your own world for the better. We only have control over ourselves, as we know ourselves. Too often, the things we don't like about ourselves dictate the most about our future actions.

This book will teach you why you do what you do, or at least give you a lot of ideas to change that. When you know why you have acted the way you have your whole life, then you can learn to do something about it. This book is about becoming something better, and how to help others become better. I will mention habits, addictions, or behaviors that can be destructive. This book isn't here to tell you you need to stop those things right now, or at all. I use them as examples of what others have struggled with. These are things that I and those around me have wanted to change about their lives.

Please don't use this book as a manual of WHAT you need to change. That is up to you. This is a book of WHY you are doing what you do, and a toolbox with ways you can change it.

Countless self-help books have been written, by extremely qualified people. Most of them take a narrow focus on a certain aspect of behavior, or philosophy, based on the narrow focus of studies of a doctor or professor or researcher. They are valuable and incredible. Very few of them tell the whole story, though, and none that I've come across can tell their readers the TRUE reasons why we all struggle with our personal problems. The root cause is always still a mystery to most people. Having strategies to improve ourselves is crucial to life on this planet. I feel that understanding what causes our problems will give us the foundation to come up with our own individual strategies to tackle our own problems. This book will give you both understanding and tactics for understanding your world and then conquering.

Too often, we look at ourselves and think we are terrible people for the mistakes we make. "Why can't I get it right?" "Why am I so messed up, when everyone around me is doing so awesome?" "What's wrong with me?!?!?!"

I hate that feeling, and I hate seeing people suffering from that same feeling. The feeling that because we want something that's unhealthy or bad for us, we must be screwed up somehow. The truth is, you are normal. Given your circumstances, your upbringing, your biology, your genes, your environment, you are probably, really normal. Whatever "normal" is.

This book you are reading has been created and crafted to distill the essences of so many of those other great ideas, and mix it up into a delicious meal of healthy self-esteem that can teach you why you are the way you are and then give you all the ideas that exist on how to be the best YOU you can be.

This book may not be revelatory or new. It may be the same old stuff you've heard your whole life. I'm alright with that. I would hope that much of this is common sense. If it is, we are on the right path, as a society. I would also hope that I've picked up something you haven't learned and sprinkled it among these pages. If nothing else, I hope I can put distinct words to thoughts you may have had since childhood, but never put a concrete form to.

I will get a little bit scientific on you, just a fair warning. It won't last long though, so if you're not a technical kind of person, I hope I explained things in a soft enough way that you can slug through it anyway, and maybe even pick up something useful as you do.

I did my best to write as simply as possible, to allow anyone of any age or education level learn from here. It may seem oversimplified for those with an advanced education, and it may seem too deep for young children. I did write this so that both ends of the spectrum would learn from it and benefit from it. This book will be "PG rated" though it will tackle some heavy, adult issues.

Throughout this book, I'll dance around the Purpose of Life, and you may figure it out as you read. By the end of this book, you'll know it, plainly and clearly. It's not what you think it is. It just isn't. You may not like it at first, but let it grow on you, and you'll realize how important it is to understand for your happiness and everyone else's. I've never found a counselor, therapist, life-coach, or self-help guru that's said it. There are plenty of people who instinctively know what it is but never thought of it in such grand terms. I've heard plenty of scientists recite it, but only as an abstract concept that didn't mean anything for themselves.

If this book is enough for you, I would love to know about it. If it serves as a stepping stone for you to get the help you do need from somewhere else, I would also love to hear about it. Find me on Facebook and send me a private message, or a post, or whatever. If I missed something important or crucial, congratulations! I'd love to know what I missed.

I've experienced a few ups and downs in life. Divorce and the loss of access to my kids for an extended period of time even after giving up on personal dreams to focus on my family. I spent a great deal of time researching more about running a business and being an entrepreneur than just about any of my colleagues only to struggle with my business and its too-many storms to handle for it to be profitable in any way for 4-5 years before ending in bankruptcy. Bad habits and addictive behaviors that contributed to a first failed marriage despite doing everything I knew how to overcome them in healthy ways. Numerous job losses and other failures that have kept me and my family in poverty and burdened by growing, insurmountable, crushing debt despite dedicating much more than the "standard workweek" to escape it and serve my employers and patients.

I still have huge amounts of inescapable student debt I haven't been able to make a dent in, let alone save for any sort of future or retirement. I haven't even been able to keep it from getting worse as interest accrues. We have no other debt than student loans, budget like crazy, and have generally followed sound Dave Ramsey financial advice the entire time. I'm balancing a new and amazingly wonderful wife, two families, a total of five kids, two or three jobs at a time, exercising regularly and staying physically healthy, and writing this book. I've also stayed active in my community, church, and profession. All without drugs, alcohol, or many of the other common addictive crutches that just about all of us depend on to get through all of this. I won't claim to be perfect in any way. I have many of my own flaws and weaknesses and still need my own escapes from the stress of life.

I've worked with therapists, support groups, and even worked as a long term facilitator of support groups. Being a little obsessed with overcoming these flaws so I can maintain what good I've found in life is the ultimate goal. Staying healthy mentally, emotionally, and physically so that I can live and keep working longer than most people live and work, is what I'm committed to. I have been gifted with a childhood that taught me how to have a strong mind and brain and a healthy body. Add to that an incredible education in the biological and health sciences, a career that focuses on keeping people healthy in any way I can while dealing with crippling fears in my patients, and a persistent fascination with all science and all religion, and you get this book.

There have been too many times where I've thought that I was worth more dead than alive. Where the pain of failure and heartache has been almost more than I can stand to live with. Times where I would go out for a walk to clear my head and not know if I was coming home at the end of it. Times where the fear of failing again have been crippling and paralyzing. There was a long time where I had to fight the temptation to step into traffic and end "it all."

When I managed to beat back that temptation, I still wished and prayed for some other "accident" to happen to me that would end my life for me. Yet for some reason or reasons, I'm still here and I've been writing this book through all of it. It is equal parts therapy for myself and a desire to lead others out of the darknesses we all face in life.

We need each other. You and I. Yes... I need you. And you need me. And you and I need friends and family... and when possible, a committed spouse or partner, and kids. We all need neighbors, and a community, a society, and a country to belong to. As much as some of us would like to be alone, to go escape off the grid, or retreat into our prepper-style nuclear-fallout-EMP-proof basement, we need to help the people around us and we need their help in return... Even if for no other reason than to have them help you build your fallout shelters (though that would give away your secret).

Humanity thrives by creating connections with each other, like a fabric. A hole or tear in the fabric in the form of a natural or societal disaster may be devastating to large areas of the fabric. As we recuperate and rebuild from these events, we can heal those tears in the fabric of humanity to become even stronger than before.

Maybe we can barely tolerate even the few people we have to interact with at work and would rather enjoy the company of our pets. Perhaps we can't stand being alone and the thought of going home to an empty apartment or house is deathly scary. So we stay up all night, often participating in questionably unhealthy pastimes, surrounding ourselves with others who need company as bad as we do.

Sometimes we like our neighbors only a few feet away. Sometimes we like them a few miles away. We all need them at whatever distance we're comfortable with, and deep down, we all want them, to some degree even if it's for the sole reason of having people that we can make money off of or use to fulfill those needs we can't meet on our own. So when our family, friends, neighbors, community, country, or other humans around us are doing stupid things that mess with our life, we want them to stop. We often force them to not be stupid anymore. We want to fix humanity. We want to fix everybody else.

The nearly universal goal of every young adult venturing out into the real world, full of ideals and big dreams for the future, is to Change the World. We see "the world" as full of problems and suffering, and we want, at some level, to alleviate that suffering, and make some meaningful impact for the better.

Around 1624, an English cleric penned the line, "No man is an island" in a book of his writings. Western society seems to like that line a lot because we still keep quoting it a few hundred years later. It means that we cannot exist by ourselves (as much as we'd like to think we can) and what we do to ourselves has an effect on others (as much as we'd like to deny that it does).

In 2010, I had a death-grip on my cell phone. It was my crutch. I couldn't live without it. I was living a 5-6 hour drive from the nearest family members, trying to survive my last year of school on my own. Not long before, within one weekend and without warning, I had gone from coming home from school to my wife and two kids each day, to being completely alone; in an apartment full of memories of them and facing an impending divorce. My cell phone quickly became my emotional lifeline to supportive family and friends. It was my connection to the outside world. At least, outside of school. I couldn't even communicate with classmates and friends easily without it. One night, I misplaced it for a few minutes and had a panic attack.

It was a rough time.

Even while still taking full-time classes in an intense graduate program, my first wife and I were just about to complete a Foster-Adoption program. In addition to our own two kids, my wife had a strong desire to adopt a child, and the San Francisco Foster-to-Adopt program seemed the best way to do it. We skipped out on family events to attend the required classes. We read a ton of books, including "How to Parent the Hurt Child," and "Parenting With Love and Logic." There were several mandatory meetings with counselors and psychologists. We were only one meeting away from completing the program and starting to look for kids to adopt through the foster program.

In meeting with these counselors, it became clear. Our life ROCKED! At the same time, as cool as I thought our life was at the time, we had some deep issues start to surface that needed addressing in our marriage. Or, through the meetings, I found and recognized some personal issues and unmet needs inside myself that needed addressing with my wife. So I started looking for help. As I tried to shed these things from myself, it scared my wife. Bad. And she left, unexpectedly and without warning. My world collapsed. The divorce was finalized in another state in less than 90 days with my kids too far away for me to see them more than once a month while I tried to finish school. My student loan debt was so high, I couldn't quit school and expect to earn enough in another career to pay back the student debt, so I had no choice but to finish. I felt so stuck and alone.

And in that loneliness, I realized how much I needed other people in my life. I'd always been proud of being self-reliant, of standing on my own two feet. Of being resourceful and independent. I realized how terribly pathetic a lifestyle that was. Without needing to depend on people outside my family, my life had been built around my life with a wife and kids. I probably put too much value on myself into my own little family. I realize now that this added to many of those issues I was now having to deal with. And my world crashed and I wished and worked and wished, and worked, and tried, to make it change. To put my world back together. I failed, pretty miserably. There's a lot more to the story than just this, of course, and that's a story for another time and place.

I learned that I couldn't change the world without changing myself and that as I changed myself, my world got better.

If the world around us is awful, we need to take a careful look inside of ourselves to find out why. If the world around us is amazing and we can't understand why anyone would think otherwise, we need to take a careful look inside ourselves and figure out why we can't understand those around us, who need our care and understanding. Since there are those who are absolutely convinced that the world is going to Hell in a Handbasket, and there are also those who are absolutely blown away in constant amazement about the world around us, that tells me that the problems with the world lie mostly in our perspectives and focus.

I know the world isn't perfect, and the natural order of things tends to be a destructive one, except for the fact that for the past 4.5 billion years (as near as we can tell by looking at the dirt and rocks around us), life on Earth has tended towards becoming more complicated and specialized. For whatever reason, an incredible combination of special circumstances has allowed life to get better and better and better at adapting and surviving. And we do this by each of us changing our own world around us. Hopefully, also, for the better.

I can't claim to know all the ways to accomplish this, but you'll find a lot of ideas in this book. I'd even be bold enough to say you won't find more of them in such a condensed, form as this book will be. You'll love it, just you keep reading and see.

A lot of us feel like we don't fit in, a lot of the time. That's pretty normal. And we won't ever always fit in for various reasons. To be unique doesn't always mean to be weird or non-conforming. Every time we try to stand out from "everyone else," all we're doing is just joining another group, maybe a smaller one. We are each unique because of the combination of ways we conform, not because we never conform. It would be impossible to be truly unique from everyone else. We are each unique because there's a mix of things we like and do that is unique to us alone. And yet we need to fit in if we want any sort of relationship. We need to adapt ourselves to fit with other people, and they should do the same for us. It's a tricky balance, and most of us are off balance, most of the time. It's the constant effort that keeps us close, that makes life work.

For some of us, life will be "worse" than others. Some of us will have it far, far worse than others. We are all born into different circumstances with different abilities, different talents, different parents and families, different environments, different educations, and different ways of thinking. There's NO WAY we can enforce any sort of equality onto every living person. Even siblings raised in the same home with the same opportunities can't hope for equality in life. And so we try to soften those inequalities a bit. That's a good thing, as long as we realize some of our inequalities are what can make humanity the coolest thing ever.

Those who are on the "low" end of life, their struggles, and the training and wisdom they can gain in the struggle, are irreplaceable and often unachievable in any other way. As Thomas Stanley pointed out in his book, "The Millionaire Next Door," there's a good reason why most millionaires are first-generation: those that learn to earn that kind of money, have a difficult time passing on all of their hard-earned lessons onto their kids. You can teach kids to earn money and how to use money, but you can't teach them fortitude and endurance without letting them fly free on their own, which most parents have a hard time doing.

For those on the "upper" end of a rich life, their abilities and talents and intelligence can be incredibly helpful to others all around them if they share it. The positions of employment they create are hugely beneficial. The products they create does improve our lives immeasurably.

Those of us on the lower end, living paycheck to paycheck while we budget and live cheap, count every penny, and still fall further and further behind: hopefully, we're teaching our kids how to work hard and manage their money and personal habits well. We're developing in them an appreciation for their work and money and small luxuries they couldn't appreciate in many other ways.

Hans Rosling, in his various talks, presentations, and his book, "Factfulness" points out that there are not "developing" and "developed" nations, there are really four levels of prosperity, and right now, most countries exist in a middle-class type of existence. And it's in great part because of the mix of help from the Upper-Class countries, and the hard work of those who've grown up in the lower class countries determined to make their home a better place. The world still has many problems that need fixing, but it has gotten incredibly better. Even better than the great majority of people believe.

I always found it wonderfully impressive that low-income families drive so much of our culture. In Junior High School in the mid-'90s, a lot of my friends started shopping at thrift stores to buy their clothes. It became really cool and trendy to have thrift store clothes or go "thrifting," finding those cheesy t-shirts, or some really eccentric pair of pants. Over the next ten years, it became harder to find cool clothes at thrift stores as more and more people took to the pastime. And then clothing manufacturers and fashion designers began selling designer clothes that mimicked those thrift store shirts. Pre-worn jeans and faded t-shirts with made-up funny logos or vacation destinations on them. 80's ringer t-shirts came back in style for a short time.

It's like people on a budget will still do what they can to remain relevant and sociable, which then becomes popular, and then high-income people begin imitating it as well. And then, low-income people try to find ways of imitating the expensive stuff, or rebel and create something new. Dr. Seuss nailed it exactly with his story about the Sneetches on the Beaches. We all want to stand out from the "other" people, but we all want to fit in, too.

It's in these gifts given to each of us —rich or poor— by the world around us that we find joy and happiness for this life. It's in growing and becoming better than we currently are that life becomes really, deeply, rich. When we are growing mentally and emotionally, then financial and physical setbacks don't keep us down forever. We find some way to deal with them, and often even overcome them, learning more in the process! And, in doing so, we change the world for the better...mostly.

What holds us back from adding to the world, from improving the human condition? Or what slows down our giving? What makes it so difficult to find good employees, or to be a good employee? What makes it so hard to find a good partner or spouse, or to be a good partner or spouse? Why do so many stupid people end up in the News? Why do we do such awful things to each other?? Why is this world so awful to so many of us?

Why is life hard? Why do bad things happen to good people? All of us feel a sense of injustice when bad things happen to good people, or when good things happen to bad people.

Even many animals have a sense of fairness and justice. There's a famous experiment done with monkeys where one monkey was given a boring food item while the monkey in the cage next door was given a delicious fruit. The first monkey sees this injustice and demands the good treat. When he's refused and turned down, he angrily throws his own food on the ground. He would rather go hungry than be treated unfairly. He is being treated unfairly through no fault of his own and he is willing to hurt himself to deal with it. How often do we do that to ourselves? When we are hurting, we hurt ourselves more!

So what does this mean? Life is hard. Life is AWFUL. Disgustingly awful. I've heard too many times at church that "God will not give us any challenge we can't handle." I understand that the purpose of this statement is to give courage and encouragement to people going through a rough time. There's a lot of wisdom behind that idea, but I feel it's still wrong. Life is more than we can handle. It always has been that way and it always will be that way. The first man and first woman who learned to overcome human nature and think and reason beyond their genetics, to learn "good from evil," the legendary Adam and Eve. The Crux of their story is that they had to deal with more than they knew how to handle in the face of deceit or misinformation, and they made a bad choice.

Just to be clear, life will be OVERWHELMING at times. It will be more than you can handle. You will drown and sink at least a couple of times in your life. Circumstances beyond your control will smash to pieces all the things you've worked hard to build. You will not be able to keep up.

Life coaches, motivational speakers, and the rich/famous/successful people who will tell you that if you just work hard enough and smart enough you can achieve what they've achieved —or something like it— are full of themselves and completely ignoring the incredible circumstances and gifts they were given to get them there, even if those gifts came in the form of crazy difficult challenges. Yet, we can still learn from them.

You need to know that you will fail, completely and totally, at times. You will somehow still be alive when you know you shouldn't be. You will see plenty of people who weren't nearly as lucky as you to survive what you did. This world is not fair. It cannot be and will not be. Most of us will try to make it a little bit fairer and that is good, but even those of us who try to make it fair will still have to deal with some of the greatest injustices imaginable.

The true nature of our character is ONLY brought out into the light when we have to deal with more than we can handle. What do we do when life gets really, really, suckie? When it is completely and totally unfair? Only by going through something like that and working our way through that muck can we begin and start to learn who we really are. It will be more than we can handle, and we'll fail miserably at handling it. We will repeat the actions of our primeval ancestors. We will have succumbed to nature. It will happen to all of us, eventually, as the universe takes back hold of us and wears our body down through its endlessly persistent aging. Entropy will win and chaos will reign, despite our best efforts. Bad things that we can't handle will happen to us no matter what. What do we do then?

If you pick up your face and stare the darkness down and decide that in the face of overwhelming defeat you are still going to keep rising up, you are on the right track. And it may take some time before you figure out how or even why you should pick yourself up in the face of all that. No one should ever feel bad for failing, and no one should feel bad for not wanting to get back up. But if there's any desire inside of you to want to get back up even if you don't know how or why you should, this is a start, and eventually, you will begin to see the true nature of yourself and those around you, and you will be pleased. My dad made all of us learn and recite often the poem "Don't Quit" by John Greenleaf Whittier which has the great line, "rest if you must, but don't you quit!" Rest if you must.

It's so easy to blame "other" people for our problems. Ridiculously easy. How easy is it, you ask? It's so easy, that almost all of us do it without even realizing it. It takes mental energy, education, and lots and lots of practice to stop doing it. For many people, it's the natural reaction to life's problems. Have you seen that Disney Movie, "Meet the Robinsons?" There's the scene when the low-life Bowler-Hat Guy, who's done nothing but grumble about his wasted life compared to the world-changing inventor Cornelius Robinson, says to young Lewis, “Hmmm, let’s see, take responsibility for my own life, or blame you? Ding ding ding ding! Blame YOU wins HANDS DOWN!”

This need to blame others usually stems from our personal problems: our weaknesses, our addictions, our compulsions, and our selfish desires. These are the killers of our relationships that only add all the more drama to our lives and cause problems among our neighbors (locally and internationally). I’m not advocating that everyone become a Mother Theresa or Gandhi. We can’t. The world would stop functioning. What we can do is use our own gifts and talents to help others in our own unique ways, from our own unique positions. And we will have to rely on others’ help, in the same way, all while retaining our own individuality. That’s tough to do.

There is a biblical commandment to love your neighbor AS yourself. Or... love others AS MUCH AS we love ourselves. The commandment isn’t to give up everything you own and possess to others who may or may not take such gifts for granted. The admonition is to treat other people as well as we treat ourselves. No better, no worse. That is a tough challenge to meet, and it takes practice, but it’s made infinitely harder when we give into pure biology, instincts, bad habits, addictions, compulsive behaviors, or personal indulgences that keep us from helping BOTH ourselves and others.

Or, even if we have that ability to help others, they may not be able to receive it because of their own weaknesses, addictions, bad relationships, and more.

Many of us have no problem thinking about our own personal needs, it’s learning to care for others that’s the challenge. Some of us, have an easy time thinking about others. You all probably know the type, the “Nice Guy” (or gal) who’s always bending over backward for others while they starve themselves, emotionally (or literally). It’s taking care of themselves that’s the hard part, and that comes with its own sets of challenges that need to be fixed.

Being in the position to help others as we help ourselves requires overcoming our own nature —our own biology. This book will talk about some of the BIG problems in human nature, but I hope that you can apply it to any aspect of your life that is keeping you from succeeding, or that is hurting your loved ones or those around you.

This book is based on a number of things, whether it's common knowledge, easy to find research, anecdotal accounts, or a healthy measure of personal experience. Any research I cite is readily available in numerous locations, and any personal stories I quote are given anonymously to protect those who told them.

It tackles the complex idea of bad habits, personal weaknesses, or "addictions." and how our biology and nature is affected by them, and how those things affect everyone around us. All those things that make us "human." If all that any of us had were only minor weaknesses and flaws, we would still have some large global issues. When we repeat those flaws through families, neighborhoods, cities, nations, and the planet, those minor personality flaws among all of us get magnified into large societal problems. Multiply even a handful of these societies and their problems together, and you get this big messy soup that is "Humanity."

Since many of our flaws, weaknesses, and bad habits are not as minor as we'd like to believe, our global and societal problems are huge. Unfortunately, we can't do much about other people's problems, we can really only fix our own problems. We can inspire others to want to change, usually by getting better ourselves, but if anyone feels like their individuality or freedom is being trounced on, they will push back, sometimes with a vengeance. We can only solve humanity's problems by working on our own. Hence, the title of this book.

I don't want this to purely be a personal improvement book or a self-help book. Nor do I want it to be an Addiction Recovery book. I want this book to teach everyone a little something more about themselves and about human nature in a way that inspires us to go help others, to love others, and help us all work together towards a better world. This is a "Society Help" book that focuses on root causes. The very basic, most simple seed to all problems in Society starts with individuals, and since we can't erase individuality from humanity, we need to understand it and understand ourselves.

I'm going to use a very broad definition of the word "addiction". You don't have to call your own problems, flaws, bad habits, or weaknesses an "addiction." You may not even consider it a problem. I had to call it something, and calling a thoughtless, compulsive, negative behavior anything else seems to downplay its seriousness. If you are free from any addictions or problems or bad habits of any kind, I hope you enjoy this book all the same and find something in it that will help you, your family, or your neighbors if they are looking to improve their lives and have asked for your help. Perhaps you'll be able to add to the knowledge in this book to help others by writing your own or joining the conversation about it.

If you have a bad habit or addiction you can't seem to kick, join the club! Quitting some of those addictions is impossible without the help of others. We need each other.

For a great while, I ignored my own addictions by not recognizing them as such. They were just bad habits or a weakness I was trying to find my way around. After trying everything I could think of to stop it, I began to suspect what I had was an addiction, or that I had a habit similar to an addiction. When I'd listen to other recovering addicts speak at school, church, or elsewhere, I could relate to them on so many levels. What kept me blind from recognizing the seriousness of my own problem was a limited definition of addiction used by most scientific organizations.

These went along the lines of "a behavioral trait or use of a substance that causes the person to suffer negative consequences, such as the loss of a job, relationship, physical health, etc." I hadn't lost any of those things from my poor choices, from my compulsive weaknesses. I hadn't lost a job, relationship, health, or educational opportunities during the practice of my addiction. I decided I must not have a true addiction, but just a bad habit.

The other thing that kept me blind from recognizing the seriousness of my own problem was convincing myself that I was clear-headed and acting normally and treating others "normally" when really, my mind was under the influence, even when I hadn't indulged in several days or weeks or even months. The problem had put a filter on my mind that kept me from recognizing its effects in my life. I thought that my behavior was "normal" and acceptable, and not hurting anyone else. Sustaining this "filter" so I could keep my life "socially acceptable" robbed so much mental energy that I can only wonder what I missed out on. A lot of the nonverbal cues, like body language, tone-of-voice, and the importance of facial expressions, were often lost on me because of either immaturity or being impaired by the energy it must've taken to keep the problem behind the curtain. In other words, the energy my mind used to keep its addictive behaviors under control in public, used up valuable resources that would have helped me excel even more than I already had, and it had also prevented me from growing, developing, and maturing in ways that I should have long before.

But no matter what I tried, how much I cared, whatever methods I used to avoid my own "addiction"—which was getting noticeably worse to me, even though I'd managed to hide it from family and friends—I couldn't stay away for more than a few months.

I'm not going to tell you what my addiction was/is. One of the most important ideas about overcoming any imperfection, any flaw, any weakness, is having the privacy to work on our own problems and only having to involve others as needed or wanted to overcome it, and most importantly, those directly affected by it. In reality, my addictions have changed forms, more than a few times over my life, and in overcoming each, I learn more and more about myself and about everyone else.

Ironically enough, I lost health and relationships, and my professional career was seriously threatened, not while I was indulging or acting out, but when I finally began my recovery. It was the attempts at recovery that threw my life out of whack.

I've seen this happen to many people. I've met doctors addicted to opiates for so long, that when they try quitting, their family members think they are STARTING to abuse. I've had friends who were only caught once they tried to escape the addiction. I don't say this to keep people from getting help. Going through those changes and getting the problem taken care of is, and was, ultimately worth it. Much like going through a painful surgery to remove a tumor or infection.

The great majority of people are still caught IN their addictions, while they are acting them out, and in many ways, that can be a harder way to achieve sobriety. Having to start recovery or self-improvement because of the coercing of family, friends, or the law, is not easy. No one likes to be told that their bad habit is dumb. Most of the time, they already know it, but they don't think it's important enough to change (or they would've done so already!), and they don't have the knowledge or help in place to replace it with something better.

By now, you're reading this and thinking that somehow you got sucked into an "Addiction Recovery" book. Yes, it is true, you'll read a lot about addictions, and bad habits. The reason why is because the problems that hold us back are the problems that hold the world back. On the flip side, the behaviors and skills that overcome our addictions and bad habits will also propel the world forward. If you're an amazing person and never have to encounter a personal demon of your own, you could still use these ideas to improve yourself into Stellar Person Mode. Or, you could use these ideas to help those around you.

I've learned that there are a great many "addictions" out there, and most are not yet recognized by society or professional organizations and that's okay. Try convincing people that Diet Sodas are addictive and bad for them! You'll get almost nothing but resistance, and even hostility in bringing it up or suggesting it. And some people addicted to diet soda will freely admit they are addicted, but there's no pressure on them to do anything about it. Addiction will mean something different to each of us.

When I describe the idea of addiction, I mean anything that we feel we should not be doing due to the destructive qualities or potential of it, but we cannot avoid it, or don't even want to avoid it, without some sort of help from others. Put simply: Something we know is wrong, but we do it anyway.

Like I said earlier, this is a broad use of the word "addiction" and some may claim that using such a broad definition can be harmful, but for those I've worked with, the broader definition allows more people to get the help they want.

That idea is what saved me. I was reading several articles about free choice, and a theme stood out to me that sometimes, a person —through events that have happened to them or choices they have made on their own— have lost their ability to choose how they are going to respond to their environment, or what's going on around them. When our more primitive biology trumps our higher reasoning skills. This is known as an addiction. This idea is the very thing that saved me. I AM an addict! I need help! And it's not ALL my fault! (though, a lot of it was, for sure) Whoa. And there are lots of other people out there kinda like me. I'm not alone. I don't have to beat myself up emotionally or psychologically, and I shouldn't do the same thing to others. I can be nice to myself AND everyone else, even though we all behave despicably sometimes. But I need help.

I went in search of that help, and in the process lost just about everything that was important to me and almost everything material. All that I retained were most of my blood-family relationships, a few friends, and my education and experience (and yes, those are worth more than anything).

What I have gained since then, I don't know that I could or would trade for anything. I've rebuilt my life almost from scratch, and while it doesn't look like much to most people right now, it seems incredible to me, knowing where it came from. I'm still in the hole, and I still am dealing with the natural consequences of my problem, and the difficult situation it has put me in, but I can deal with it with a clear head and a clear conscience—most of the time—and to me, that is like having a superpower. The confidence I have, the emotional intelligence I gained, knowing what I know and feeling the way I feel, is the greatest source of strength I have ever known. The experience and knowledge I have gained and shared along the way is phenomenal.

As a kid, I LOVED science. I loved astronomy. Learning about the stars, blackholes, quasars, pulsars, nebulas, planets, everything. I was fascinated with nature. With insects and oceans, and plate tectonics. I remember as a ten-year old proudly pointing out to my family at dinner my self-made observation that animals have a lot of the same bones we do!

Near the end of college, I had a moment of revelation. I'd taken physiology and anatomy, as well as chemistry and molecular biology. During a Biochemistry lecture, I had this moment in class where I realized I could picture in my head and make sense of everything from the smallest particles we know of, all the way to the largest galaxies and clusters in the universe!

Since then, I've also studied psychology, religion, history, and a whole lot of human nature. I've had many similar moments where I've felt I could tie in all of human existence from conception and innocence to world wars and genocides, and make sense of all of it, or all of it I know of. This book is an attempt at capturing a piece of that.

This book is a big-picture kind of book. There will be many specifics discussed, but for the most part, it is a grand landscape of the scenery of what holds us back, or those unwanted behaviors, those things that we do that hurt ourselves, those around us, and the society we belong to. Whether you decide your problem is an addiction is up to you. There's no defining point that will be the same from person to person. To me, an addiction is something harmful to myself or others that I can't stop doing by myself. On the outside, it looked like a bad habit, but I knew I couldn't stop on my own and I had tried for years.

I'm writing this book so that you can understand both yourself a little better, and everyone around you a little better. Especially when people are acting irrationally.

I would argue that we all think we act "rationally" all the time, or really, almost all the time. It may not seem rational to other people, and the general consensus in society might be that a person was not behaving rationally in a given moment, but to that person in that moment, what they are doing makes perfect sense to them at that moment in time because of 1) the information they had 2) the functioning of their mind, and 3) what they could do about the situation.

All three of those things are planted in us by our circumstances, our thoughts, our actions, our upbringing, our hormones, our own peculiar fears and addictions, our physical health, our emotions, the people surrounding us, and our culture throughout our life. There is some ability to make choice mixed into that —and it's incredibly important!— but I would argue that it's actually a relatively small amount of "choosing" that actually happens. Most of our "choices" —or what we think are our choices— are the easy, no-brainers. Really, our emotions and our intelligence choose for us, with an answer that seems like the obvious choice and then we take the credit for being so wise.

Our own universe is our own universe, and we are at the center of it. It will be at least a bit different universe than anyone else's. And the more we can understand our own and someone else's actions and thoughts and feelings, the more we can make our own universe and their universe just a bit more like the real one. Understanding ourselves and each other will do more to build humanity and society, than anything else. This book will teach you to understand others.

Humanity has thrived because of its ability to cooperate within itself and with its surroundings while preserving individuality. Humanity is the most capable life form this planet has seen. No other single species has been able to do so much with the dirt and water around us than humanity has. This ability only comes because we can work together, share our knowledge, and help each other while promoting and preserving individual gifts and experiences. Working together involves avoiding destructive behaviors while holding onto and promoting constructive and productive behaviors.

The circumstances, philosophies, and behaviors that harm us as a society, as a species, as a culture do so by either dispersing large groups of us, fragmenting our communities and over-valuing individuality. We are also harmed by going too far the other way and eliminating the value of an individual and over-valuing the community. Both, the community and the individual, have to exist together, and our brains are wired to do exactly that... most of the time. Anything that causes us to look at other people as being slightly sick or twisted only makes us slightly more sick and twisted ourselves. Anything that compels us to look at another group of people of any kind with disdain of any kind, is ultimately bad.

We have to gain knowledge, we have to learn, and we have to gain experience for ourselves and learn of the experiences of others and put it all together as best as we can if we want to make this world a better place.

This isn't going to be a religious book, though I will reference religious ideas throughout it so that we can all understand each other a bit better and recognize that most of us are working towards the same end, to make the world, and ourselves, a better place.

There's a common argument that without the hope for Heaven, or the threat of a God, why would we be altruistic? Why would we want to improve ourselves or help other people? What would be the point? Why use our own energy for someone else? Why make sacrifices of ourselves for others? Unless there's a reward for being "good," why be good? Biology and science show just as many reasons for this as religion does and I'm fine with that.

There's an argument that since we can never be truly satisfied with life, we must be meant for another, better life. While I like the idea, the truth is, that if we could be satisfied with life, we would stop living, or at least stop doing....anything. Never being fully satisfied is what keeps us alive and growing and working, either for ourselves or for others.

While I may reference religious ideas and principles in this book, I'm not going to try to convince you to believe the same way. This isn't the book for that. You don't have to believe in God to be a good person, and I don't expect you to believe in God just because I said to. Being Good? Yes, I hope to help you be more of that.

Where we find "God", we also usually find rational, scientific reasons for something to exist, and where we find rationality, we also often find reasons to believe in God. To me, the universe is inseparable from the God who's represented by it. Where we find rationality we also find God, or Goodness, and vice versa. Because of this philosophy, I try to look for the rational reasons why faith and religions have their peculiarities and unique doctrines and beliefs, and I also see the miracle of life as something that makes perfect scientific sense in this enormous universe. It is incredibly rare and mind-blowingly amazing at the same time, and we get to be a part of it, however it came to be.

2

Background...or...Why Do Bad Things Begin

Before I address that header, I'm going to give a grand summary of life on this Earth. There are three guiding principles to ALL life on this planet:

1. Pass along your genetic material, your DNA, to the next generation
2. Take what you need from the world to accomplish Rule 1.
3. Do numbers 1 & 2 better than others or go extinct.

All of life, as we know it, boils down to those three things. I'm not going to discuss what happens before or after life, extra-terrestrial life, or what the meaning of life is for religious or philosophically-minded folks. This is straight biology. If there were or were not a God, then life, or nature, as we know it now, would still complete those three ends... as far as we can tell. Have babies, eat the right foods, and live long enough to make both happen successfully.

From the simplest of viruses to bacteria to mushrooms, to mammals, to homo sapiens, biology is hard-wired to produce offspring, and accomplish whatever is necessary to do that. How this is accomplished varies as far as the imagination can go and beyond. If you can think it, there's probably an organism or life form on this planet that does it. DNA is this crazy stuff that is possessed entirely with replicating itself, at just about any cost. DNA will do just about whatever it takes to make a copy of itself, and that, sometimes, is what gets us into trouble more than just about anything else.

For us humans, that means, find someone else to reproduce with and find some way to convince them to reproduce with you. If you can't manage those two things, well, your brain will just go ahead and find some substitute that'll either take away the stress of failing those things, or trick your DNA and the sexual parts of your brain into thinking that you're at least using your reproductive organs and hormones in some way. There are untold ways of accomplishing and satisfying those urges or sedating those impulses, or, turning down the volume on the failures of our life by using substances and behaviors.

Life and DNA (they're basically the same thing) have figured out a whole lot of ways to accomplish this, across all species. Sex is a pretty dominant method DNA uses for perpetuating itself. It works really well.

Some species have no gender, some species have both genders mixed into the same...thing (most germs, much of sea life, fungi, some flowers, and trees, for example), and some have very separate and distinct genders (peacocks, anyone?). Some animals —like insects, turtles, and frogs— leave hundreds of eggs alone in the wild to fend for themselves in the odds that a few will survive. Some animals —like elephants, bears, and primates— have only a select few babies that are incredibly protected and nurtured to adulthood to ensure their survival.

The end result is really as the fictional Dr. Ian Malcolm says in Jurassic Park, "Life...finds a way." In humans, it's pretty crazy. A great many men are ready to pass on their DNA just about all the time, while a lot of women, don't feel that same push except while they're ovulating. With that disconnect, everyone gets really creative trying to bridge that gap in sexual desire and push, while trying to maintain civility.

I'll mention in a bit why some of these creative attempts arise, but the simple truth is that men are constantly, constantly getting pushed towards having sex of some kind by their body, their mind, and just about everything in them. "Civilized" men learn how to restrain it, keep it calm, and under control, at least in public. Being in a steady relationship often helps a great deal to keep the pressure from building. Finding some sort of outlet is helpful, so is reclusion. It's slightly easier to resist temptation when you're not staring it in the face every day.

What does this have to do with addiction? I'll tell you and then get back to the biology. Our addiction is driven by biology, pure and simple. "It's in my genes." Or, "It's just who I am," "It's their fault. I had to do it to survive!" are commonly touted excuses or rationalizations to participate in whatever destructive behavior we feel like doing: Stealing, lying, cheating, angry outbursts, etc. These statements about genes and biology are true, to a degree, but there's more to the story. Our survival instincts are ingrained in our genes, and addictions hijack those survival instincts. BUT (and that's a BIG BUT!) we humans have a unique brain that can learn to override our survival instincts and our addictions.

First, let me get back to what makes us, us.

Our DNA carries instructions for making all the machinery in our body. DNA works like this: Imagine crinkling a bit of foil or spring into whatever random shape it wants to form on its own. Do that a billion different times. After a while, you'll get a few odds and ends that got crinkled into useful shapes, like the shape of a spear, or a cooking pot, or a gadget that can break specific rocks apart, or something else. The crinkling action is done by DNA, and the "foil" or "tools" it makes are your hormones, proteins, and enzymes; the machinery and structure of your body. All of DNA is the blueprints for proteins that form their own crinkled blobs the way nature tells them to. Just like with the foil, whatever DNA forms useful bits, or whoever is good at crinkling foil into useful shapes, gets to gather and eat food and then have babies. DNA that makes useful machinery gets to make copies of itself, make babies, and get passed onto the next generation.

Large bits of DNA that make up a protein machine are called genes and they get turned on and off at various times in our life by all sorts of different ways. There's one gene for a specific hormone that gives our brains an extra boost while we are still in the womb. It turns out that this gene is almost identical to our nearest genetic relative, the chimpanzee, except that for subtle, chemical reasons, ours stays on twice as long as theirs. It's one of the few key differences between us and chimps and makes humanity special.

Contrary to popular belief, we do not all start out as females. Our organs start out gender-neutral, and if a specific step doesn't happen, they'll almost always automatically turn into female organs. There's a gene on every Y chromosome that has the sole purpose of turning a baby's sex organs into testes. It does nothing else, ever. It turns on once, for a few hours when a male embryo is just a couple of weeks old, and then it hibernates and sleeps for the rest of that person's life. Once in a great while, that bit of the Y chromosome will get mixed up in the repackaging that DNA goes through and may get stuck onto an X chromosome. That's one way how mutations happen. If a mutation helps life or is an easy mistake to make, it will stick around. If it doesn't, it'll stay a rare and random mutation, or eventually disappear.

Now, there's a whole lot of other things that can mess with this gender DNA process before, during, and after birth and fortunately, those genetic mutations are rare. We'll all have various mixes of masculinity and femininity to our personalities because of our upbringing, our environment, and our choices, regardless of our genes. Our brain and what happens to it will be the major determinant of our life.

As our brain goes, so goes our body. Our brain is the master controller organ of our body. It makes all final decisions. Whichever genes and chemicals promote brain health, also promote our life and functionality and survival. These are also the genes and chemicals that get hijacked by addictive substances and behaviors. Interestingly, the plants that produce these addictive behaviors get to stick around, because we like them. Our brain says to us, "This plant makes me feel good, so let's help this plant grow," and the plant thrives like Marijuana does. Sneaky plants.

We, as highly-developed, highly-complicated homo sapiens—humans—have an incredible amount of DNA for doing an incredible amount of things in our body, and a lot of them contribute to the end result of preserving and passing on our genes. We do this by cooperating with other people in large complex societies, and by listening to our own individual hormones and instincts.

When your stomach is empty, it sends little hormonal text messages through your blood, into your brain and "ding" you need a Big Mac. If you don't get that Big Mac soon, you're gonna start going all crazy on people until you get a Big Mac. You become hangry. True story. If your experience has taught you that eating a Big Mac would make you feel worse in the long run you might instead go eat soybeans or a protein drink, or oatmeal, or whatever you've learned satisfies your cravings at that time in your life. You can also learn to recognize "hanger" in yourself and compensate for it so that you're not taking it out on those around you.

It sparks such a broad, mass, signal, to your brain that we call these kinds of sweeping signals "emotions." Hunger and food are enormous triggers for emotion because they are so important for survival.

Romantic love and family love are incredibly potent triggers of emotion because, literally, your entire brain is designed to accomplish family love. Your entire brain is required to form a family, have children, and then raise those children in ways that help them be healthy and have families of their own. And yes, we do use all of our brains, not just 10% or whatever cool number it is people say. It makes for a believable movie-explanation for the difference between us and superheroes, but, sorry, just not true.

Without some congenital deformity, drug abuse, or traumatic injury, we all use all of our brain. When we are tired, scared, angry, a bit hormonal, not feeling well, trying to hide some lie or guilt, or we're a little loopy because of some drug or medication we're taking, then our brain isn't working 100% and our ability to make good decisions suffers. Too many relationships end when our brain isn't working right.

Keeping our brain sober, not just from drugs and alcohol, but also from out-of-control emotions or unbridled hormones, is the most important skill we can learn in life. Our survival instinct can often get shut down or destroyed when our brain is under the influence of something. Destin Sandlin, on his Youtube Channel, "Smarter Every Day" experiments on himself showing this. He had the opportunity to visit NASA's hypoxic training chamber and see what happens when your brain doesn't get enough oxygen.

The doctor on duty let him keep his oxygen mask off until his oxygen blood levels dropped to dangerously low levels. At that point, all Destin could do is give a nervous grin and simple answers. He couldn't think well enough to put his mask back on even though he looked conscious and awake. Others in the chamber had to put his mask back on for him. He couldn't have saved himself no matter how much he wanted to, his brain wouldn't let him, impaired as it was. If our brain has a hard time taking care of ourselves under any sort of impairing influence, it will not be able to effectively take care of others as well as we should. By the way, you should definitely check out Destin's channel and watch that episode, #157 "Why You Should Put Your Mask On First (My brain without oxygen)."

Cooperation between people—and groups of people—is incredibly helpful and efficient because it helps us survive and thrive! We compensate for each others shortcomings and weaknesses. The genes that promote the parts of our brain that help us get along get passed on better, and so, our society should continually improve.

Our ability to move and act and do things is incredibly useful. Our ability to think and reason is amazing! The parts of our DNA that help us with those things, get to stick around, usually.

Some genes are complicated. They can be good and bad, helpful and destructive at the same time. Aggressive Alpha-type personality is very genetic and incredibly effective at getting stuff done and getting people to work together. The flip-side is that those aggressive personalities can also wreak havoc on societies, communities, and families if unrestrained or untrained. Most of our personality starts off genetic but can be altered hugely by our environment.

All life forms, however ancient they are, have cocktails of genes and hormones trying to control and influence behavior. For our simple bacterial friends, their life-goals are super simple: soak up nutrients in your immediate vicinity and when you've got enough, copy your DNA and split into two. Done! Voila! Rinse and repeat.

For higher animals, there's hunting and herding and growing of food, and mating rituals like the displaying of prominent tail feathers and physical prowess, or buying of flowers and chocolates and romantic dinners. Then, koo-koo-kachoo, once babies are on their way, there's instincts and ingrained behaviors that push the mothers into caring for their young: Instincts that cause ducks to watch out for their ducklings and for those ducklings to follow their mommas, and instincts that make momma bears crazy-protective of their cubs. And for every animal on the planet, there is no reason to ignore or not listen to these hormones.

In fact, I would argue that they CANNOT ignore these hormones. They are just a crazy-complex biological robot doing what their DNA and the environment tells them to do. I watched a show about dragonflies. Their eyes directly control their wings. There's no reasoning or thinking that goes on between what they see as potential food and what they attack. There's no rationalizing, no feeling. Their eyes pick up an object that is the right size for them to catch whether it's an actual fly, or a dot on a screen, and they automatically fly at it like a guided missile. And this system works so well that they have the most successful "kill rate" of any creature we know. They are simply a very complex, reflexive and instinctive machine.

If we keep human children from most of the normal triggers that spark our sex instincts (viewing sexual acts, nudity, seductive behavior, talking about it, etc), we can keep them from being interested in, or thinking much about their sex organs, until they go through puberty. After that, it's a different story. While human brains don't finish growing and developing until between the ages of 25-35 years old, human sex organs are ready to work during early teenage years. Sex drives typically wake up and begin to work on their own between the beginning of puberty and adulthood. Some people won't feel or notice their drives working until they're in college, some will feel it in junior high school, or even as early as elementary school.

Learning to control those drives when they can barely control their emotions and bodies, is a tricky time for most people. It's also the time that most people begin to look for distractions, escapes, or ways to satisfy their instincts. It's a messy, awkward time for lots of good reasons. With hormones, triggers, stresses, responsibilities, and pressures from society and friends all pushing and pulling on us in so many different ways, it's a wonder we survive it emotionally intact at all!

The environmental triggers and cues that spark and flare up our sex-drives can be very destructive to teenagers. The emotional trauma that can come from a sexual experience at too young of an age is so devastating and hard to overcome that many societies have placed very strict rules and laws regarding exposing or pushing intimate behaviors on kids and teenagers who are not emotionally ready for it. Just like how addictions are much more easily created in younger people, co-dependencies created by forming romantic relationships while too young can interrupt our ability to form healthy attachments and connections to others as adults.

We can't and shouldn't ignore or minimize the power of environmental cues and triggers on our or anyone's behavior.

The Youtube Channel, "Great Big Story" does a segment on the vocal actress, Tara Strong, who did the voice-over for baby Dill Pickle in the Rugrats cartoon movie. The producers interrupted her while she was recording to tell her that her baby sounds were making a new mom in the studio start lactating. The sound of a fake baby made by a stranger made someone else start leaking milk unconsciously.

Addictions, despair, and destructive habits form when one of these instincts is hijacked, mimicked, or copied by some substance, or behavior, or traumatic experience. If we were stuck with what the environment throws at us, we'd be sunk if we were unlucky enough to experience something that takes over our brain. Somewhere, once upon a time, in a land far, far away, humans developed the gift of being able to override their hormonal biology.

They learned to control their instincts and drives by using their newer outer-layers of their big brains. But this control is easily lost because of our strong emotions, faulty memories, and traumatic experiences. It's a very close battle between our inner, primitive, reptilian brain against our advanced, civilized, refined and cultured human brain.

Our emotions, yes emotions, are a product of our biology. Happiness, sadness, fear, anger, joy, amusement, pain. They're all brain-wide responses to the environment. These are not human-only artifacts, they're found in species from fish to birds, to mammals, and especially primates, and they are all used for survival at some level. Go look around the internet for a bit and check out how many videos people have found of animals doing caring and gentle things for other people or animals, or of acting very human-like in their emotion.

After watching some of those, it's easy to think that animals act unselfishly, out of the goodness of their big hearts. Except that they are every bit as concerned with their own survival as we are and it shows very readily when our life or their life is threatened. The reason we avoid wild animals is that we can't know when they are going to be angry at us or hungry for us. Even if you see video clips of people swimming with Great White Sharks, to bring awareness to such amazing creatures, know that they have made sure those sharks have been well-fed before attempting to swim with them outside a cage.

Nurturing, caring, and altruism are just as genetic and biological as hunger and sex are. We see people with pet lions, tigers, and bears, and other exotic and normally dangerous creatures. It's amazing to watch owners play and wrestle with these incredible animals. Unfortunately, it doesn't often take much of a trigger to wake up their normal instincts, even just for a few quick moments, and they will do something incredibly harmful or fatal to their owners.

Our bodies are squirrelly things. A quick Google search, and even less life experience, will show more stories than we want to see of men and women ruining families and marriages while under some hormonal stress, whether its a sexual urge or a PMS-driven outburst.

Though we don't like to discuss it, there's more than enough newspaper articles advocating marriage counseling, blog post confessionals, or Quora questions and answers about that topic.A friend told me about her husband, "There are some days I can't imagine any good reason why I put up with him.

He just can't do anything right! Then, the next day and most days, he's the Hero of my world and I can't even remember why I was so angry at him!"

Similarly, there's plenty of men ruining relationships trying to get more sex or eroticism out of life than they can get from their dedicated partner. "You're as cold as ice" sang the members of the band, Foreigner, back in 1977, and the song is still being heavily played on classic rock stations today. Even if they have a healthy sex life with their wife, a lot of men are pushed by this unseen force to go search out more, even when they can readily recite all of the bad reasons they shouldn't. WHY? I'm glad you asked. I'll explain later.

True, the husbands and boyfriends of PMS'ing women sometimes add to the problem through ignorance or their own obnoxious shortcomings. And sometimes the wives and girlfriends of sex-hungry guys can be more distant and withholding than they should be for a healthy relationship to function. Perhaps they, too, are acting out of ignorance, immaturity, unresolved trauma or something else. In the end, everyone has to tackle the challenge of living with your hormones, taking responsibility for them, and finding healthy ways to get through life with your families and relationships intact if you can.

Why should you put up with this difficulty? Why should you put in the work? It will be different for each of you. Maybe it will be so that you can gain a greater sense of achievement when you hit your 50th wedding anniversary, or so that your kids learn the good behaviors that will carry them into adulthood and into their own families, or to fulfill your own personal emotional and spiritual goals.

This world will function better when we learn to master ourselves, especially that most difficult task of mastering our sex-hormones: those hormones that are so powerful that life has persisted for over three billion years on this battered and bruised wet rock exposed to the dangers of the universe, and each other. It takes a collective knowledge of all of human history and learning to figure out how to get our bodies and minds under control of something more than pure biology.

It seems anything with a brain experiences some level of emotion. All creatures take joy in eating and reproducing. We're all just following biology. I've seen more than one cheesy episode of Star Trek where humans encounter some alien life form that doesn't understand Emotions like we do, and the glowing moral of the show is how awesome we humans are for having them. I would be surprised if we ever find an animal or creature anywhere in this universe that doesn't rely on their passions or "emotions" for survival and success.

These emotions are a mix of instinct, drive, reflex, and a tiny bit of intentional reasoning. At the top of the evolutionary ladder is us, humans. But how much control do we have over our own selves? Whether we have some sort of control over our instincts and drives is very much up for debate. Do we have "free will"? Are we masters of our universe? Can we actually choose something against our DNA?

There's no definitive answer from anatomy, biology, or psychology, which kind of makes us unique among the biosphere and leaves it a mystery. Or we just have further to go in our research. But for now, we have to accept the fact that we have more to learn and we need to take responsibility for the great majority of our actions, instincts, and reflexes.

What's the difference between a "reflex," an "instinct," and a "drive"? There's a healthy debate among scientists, psychologists, philosophers and anyone else who cares. I don't think it matters what the difference is, except that some of our instincts are more complicated than others, some are less complicated. So perhaps "drive," like the Drive for hunger, sex, or wealth is a better word. For some, an "instinct" is automatic behavior and does not need to be taught, like a baby rooting to find it's mother's milk, or males rutting to show off their physical prowess and impress the ladies. Reflexes are simple and fast, like what you do when your finger touches a hot pan.

Some neat instincts are like how Momma Turkeys will care for and nurture ANYTHING that makes the cheeping sound of a baby turkey. Put a baby weasel in a turkey nest and play a baby turkey recording, and the momma turkey will take care of it! Male robins will attack anything that slightly resembles another male robin in its territory. If a male robin spots a badminton birdie with red feathers laying on the ground, it will probably attack.

Could you find some way to make a mother turkey or male robin resist those instincts? I'm sure you could, but it takes some sort of powerful stimulus to override an instinct like that. Humans don't need quite such strong tactics to override our instincts, usually.

Instincts and drives can be hijacked by improper environments and experiences. Ducks and geese can be "imprinted" and convinced that a pair of boots are their mothers and follow whoever is actually wearing those boots.

Some of us are imprinted to hate country music and despise it with a passion. To others, it is the only pure music that exists. To some people, cottage cheese is eaten with salt and pepper. For others, cottage cheese is eaten with diced fruit. If you get those two different groups of people together, you might end up with some fireworks and incredulity. "You're eating it WRONG! That is so weird! Oh my gosh, I just threw up in my mouth a little, just thinking about it!"

Another example, Hawaiian Pizza. Best pizza ever, or worst crime against humanity invented. Much of it has to do with how we were raised and our first imprints and emotional experiences with it.

And if we are this passionate about a few food items, we can get CRAZY about things we truly do care about.

When humans are initially sexualized or feel those first associations with sex while being exposed to, engaging in, or interested in some behavior, that circumstance leaves a lasting impression that can stick throughout life. What also tends to happen is that we seek out the experiences that imprinted on us and we reinforce that imprint stronger and stronger over time. It become self-reinforcing and we can get stuck into a lifestyle, habit, or desire that we may not even like. But we need it to fill that hole that it fills. This can happen with any habit, drug, fetish, desire, or coping behavior.

Some of us will get a buzz while taking a prescription pain med after surgery, or during our first beer, and something about that buzz will stick with us like a scar that we can't get rid of without serious intervention. We have become "imprinted" by the behavior or substance. We should be careful what kind of imprinting we impose on our children and ourselves. When two people who have been imprinted to different ideas about sex try to make a marriage work, it takes a lot of bending or give and take, from both sides to come together and make that marriage work.

On the flipside, without some imprinting, many animals raised in captivity will ignore or reject their own children (if they even bother to have them) since they were never taught by example from their own mothers and fathers, or because their environment is foreign enough from their natural habitat that the right cues, triggers, and hormones, don't become activated to push the parents into mating or raising their children.

We like to think that certain instincts are ALWAYS there. We also like to think that because something feels "instinctive" it MUST be part of our genes or it must be helpful or good. Both ideas are wrong.

This is where addiction (or whatever you want to call it) comes in and takes over. We have an incredibly complicated brain that depends on chemical signals from other organs, from the food we eat, from the environment we interact with, and from special nerve clumps turned on by those things. When some new experience triggers some mousetrap in our brain that is normally used for survival, our brain associates that experience WITH survival and our drive gets hijacked to seek out that thing, often at any and all costs.

This hijacking can override all other survival instincts if it continues long enough and is strong and potent enough. Not being able to take care of yourself, or trying to satisfy some uncontrollable craving at the expense of yourself or the neglect or abuse of others can cause huge problems to yourself and our community. It's not surprising that isolated and primitive tribes of people can take such harsh attitudes towards non-conformity. Even today, within families, religious congregations, political parties, schools and universities, and neighborhoods, non-conformity is as threatening, by nature. It takes a small amount of conscious or practiced effort to look past any differences and think outside the box. Within companies and teams, conformity ensures survival. The same is true all across humanity. So when someone breaks from the mold, our automatic reflex is apprehension and trepidation, until we understand why that person is behaving peculiarly. It can often turn to admiration very quickly. Sometimes non-conformity is dangerous, sometimes it is helpful.

No matter what, overcoming our instincts and drives is incredibly important to the advancement of humanity on the whole. Sometimes we fight against our instincts in rebellion against established norms. Sometimes we fight against our instincts to become unified.

Many people fear the "Singularity": that moment when computers, robots, or machines will gain enough consciousness and awareness and ability to act for themselves as they see fit, and take over humanity. Countless movies and novels have been written on this idea, from Skynet in the Terminator to HAL in 2001: A Space Odyssey. From Johnny 5 in Short Circuit to iRobot.

Elon Musk, that amazing entrepreneur who's done a lot to promote and create so many amazing technologies has spoken openly about his concerns regarding artificial intelligence. Isaac Asimov was writing about it over 100 years ago when the most advanced machines he knew were nothing more than gears and cables!

Instead of fearing when we create artificial intelligence, we should be still working on overcoming our own instincts and drives. Not to get rid of them, but to learn how to teach the world to be in charge of them. Something we should look forward to is when we've gained enough knowledge and experience as a species that we finally learn how to overcome our own biology and genetic programming to act purely for ourselves as we see fit. That will be an amazing day when we are able to bestow that gift onto every new generation born to this planet.

To accomplish this, we will ALL need to know more about our bodies than we currently do and to help with this, I've compiled information and resources from numerous disciplines, sciences, and fields of study, across the spectrum, to help you: The Reader. It's also helped me, immeasurably, getting it all together for you.

Many religions of the world have as their aim, the goal of overcoming "the flesh", or learning how to be Master and Controller of their body, instead of a slave to hormones, physical desires, or emotional impulses. Whether it's Buddhist monks working on self-mastery to achieve Enlightenment, Hindus learning through the many chances of reincarnation to get life right so they can enter Nirvana, or Christians, following Christ as their exemplar into subjecting their flesh to the Will of God. In this case, God would be that force of goodness that is higher than nature.

Our goal is to advance beyond biology. To become Super Natural. To rise above our normal selves. To Improve. At some level, we all want to escape the boundaries of "life" and become something more. To leave a legacy.

There is something inherent in humanity, across all cultures and geographies that is inspiring us to be better than we are, and to somehow conquer and subdue our physical bodies, not unlike the way a horse trainer has to break a horse to be ready for a rider. A rider who is smarter and wiser than the horse is, who can use the horse's strength, speed, and endurance to accomplish much more than the rider could do alone.

Crazy enough, our genes and our biology, are giving us the ability to do this in a way no other life form has even come close to. Most of us won't achieve this mastery within our short mortal lives. But it's within our sights and it's close enough to reach for that large numbers of humanity keep trying anyway. And in that individual reaching, we push the world to get better, one person at a time.

A person whose instinct for cooperation and self-preservation has been damaged can often become a danger to the group, especially in a small isolated group. As our groups and communities get larger and stronger, we start to see the value in our unique and peculiar individuals and the ideas they bring with them. We can start accommodating for their destructive tendencies in exchange for their creativity or other productive and valuable skills. In the end, a large, robust society is benefited by each person's strengths and weaknesses.

Interestingly, self-destructive behaviors can get bigger in large societies. Instead of just an individual behaving in self-harm, we find large groups of people engaging in societal self-harm, in just as many forms and styles as there are individuals. Whether it's ancient infanticide, or the persecution of another group based on gender, race, geography, success, lack of success, political leaning, clothing style, music preference, or whatever.

If we don't like someone else who hasn't actually hurt us, we are engaging in societal self-harm. We are "cutting" ourselves and we may not even realize it because it always seems justified to ourselves. An individual who cuts themselves can usually come up with some justification or reason for doing so, a large group can come up with reasons for their bad behavior even easier.

Our instincts and drives are valuable and powerful, overwhelmingly so, but with help, time, experience, knowledge, practice, and sometimes medication, instincts that have turned into addictions can be beaten. We, humans, have a huge brain among the animal kingdom and the whole front-outer section of our brain is designed to help us override our more primitive instincts. In many addicts, the connections between that frontal cortex and the rest of the brain have been damaged or weakened to some degree.

Once those connections between the checks and balances of our brain are weakened or damaged, our animalistic inner-middle brain usually takes control and doesn't let go of it easily. Like a military commander who launches a coup against the government. Sure, it lets the outside of our brain have control when it's NOT that important, but when it counts? It says, "I'm in charge."

It lets us play at our fun little games of work and family, but in the quiet moments, often when nobody is watching or caring what we do, it has full rein. If it gets used to such free rein, it starts to stay in control during all aspects of our life, wreaking havoc on our relationships and obligations.

Our brain acts like the helicopter parent that hasn't learned to trust his or her child or a codependent spouse who can't really let their partner be themselves for fear of embarrassment.

We tell ourselves, "I'm in control of myself," when really, what we mean is, "I can act in a sociable way most of the time." Or, "My teenager can decide for herself" while we breathe down her neck or guilt her into conforming. The need for our animal brain to control us, or others, is very strong, but it can be overcome.

We have a "reptilian" or "animal" inner brain whose sole job is to ensure our survival at any and all costs. It doesn't use rationality. It doesn't think. It simply reacts, mostly to threats, and to a lesser degree to rewards. We all inherited a brain that is more likely to be afraid of losing something we love than of taking a risk to gain more. We also have a super-advanced outer-brain that, if given enough knowledge and practice and memory, can override the inner brain. It can use rationality, thoughts, feelings, and emotions in weird, creative, complicated, and new ways to not just survive but THRIVE in a complicated society and world.

But! it takes a lot of practice, time, and repetition to build that outer brain up, like a muscle, or like learning a new skill. And fortunately, except in cases of certain, rare, types of brain damage or disease, our brains can learn and change and grow through our entire life. Not unlike a form of government where the power is given to the masses, rather than a minority ruling class that has control of the military or wealth.

Luckily for everyone, especially for those who've ever experienced REAL, traumatic, brain damage of any kind, our brain is "plastic," or, it can still change and grow and repair throughout our life. Old dogs CAN learn new tricks, and addictions can be overcome, though it may be the hardest thing you've ever done.

You're gonna see this phrase, "the hardest thing you've ever done," come up a few times in this book, they all apply to this same thing and is meant to give you reassurance that it will be hard, but it is do-able. Difficulty in working with or resolving your feelings about some "natural" tendencies is expected and normal. Going against your genetics and environment are going to feel impossible, even though they are not.

3

How Does the Brain Work—Can We Make Sense of Nonsense?

This will hopefully be the driest chapter in the book, so let's get it out of the way early on. I did my best to keep it easy. This is some nitty-gritty of brain function, and by the time you finish it, you'll officially be an expert in neurotransmitters and mind control. Officially.

No. not really, but you'll understand a bit about how the foods and substances you eat have such a crazy-profound impact on your behavior.

Your brain controls the rest of your body. End of story. Romance and spirituality talk about your Heart like it's the center of feeling, but, I'm gonna ruin all of that for you, your brain controls your heart, and interprets all those warm-fuzzies you are feeling in your chest, into the thoughts you are going to have.

The heart has its own beat that'll work without any input from the brain, but that's about all it can do by itself. It only has one setting on its own and that setting is, chill-slow. The brain does EVERYTHING else, we think. That burning you feel when you see someone you love? Brain. That anxiety you feel in your stomach when something scary is about to happen? Brain. That overwhelming rage you feel when someone disses yo momma? Brain.

What you feel in your heart is often related to the Vagus Nerve. Vagus means "wanderer" because this major nerve wanders from your brain to your heart and stomach, having big effects and changes along its way. It's very involved in your feelings of safety, relaxation, love, and your "rest and digest" system. This is the opposite of your "fight and flight" system. It slows your heart rate down, reduces stress and inflammation, stimulates a healthy immune response, and a whole lot more.

There's some evidence that the things that stimulate your facial nerve (your cheeks, lips, tongue, and teeth) have an effect on the vagus nerve. We naturally turn to things that stimulate our face, especially around our mouth and throat, to reduce stress and anxiety. This can show up as clenching and grinding teeth, eating, smiling, and being friendly and intimate with others. All of this is an electric response from nerves to the brain and all of it's chemical and electrical machinery.

As if the brain wasn't enough of a manipulative, controlling, dominating half of the relationship, there are the pituitary glands (they are a pair) sitting right in the middle of your skull, wrapped in a saddle of bone, safe and protected and sheltered from just about everything. It cranks out tiny droplet-sized recipes of hormones that do all the dirty work of the brain.

When the brain isn't going to do something directly through the electric signals of the nerves to muscles, it usually calls on the pituitary gland. The pituitary is a little sack of brain tissue dangling precariously from the lower middle of the mind and is so fragile, delicate, and supremely important, the body has evolved our skulls to wrap it in bone. It is responsible for producing and controlling Growth Hormone, water-balance hormones, sex hormones, metabolism, blood pressure, feelings of love and attachment and more and more.

The little pea-sized nugget wafts a few molecules of oxytocin into your bloodstream and suddenly, you're in labor ready to start pushing that 9-pound goliath out of your uterus (if you have a uterus, with a baby in it)! It oozes a little bit of Thyroid Stimulating Hormone (TSH), and your thyroid responds by releasing some stores of iodine-cool-stuff, and your metabolism kicks into high gear burning some fat reserves. You eat some salty food and the pituitary gland squirts a bit of ADH into your bloodstream, telling your kidneys to get their rear in gear and get back to work and start pulling the extra salt out of your blood and into your pee so you can get rid of it.

What if your kidneys are already busy working because you just finished that bag of baked potato chips? They send a message back to the brain telling them to get off their case, and why don't they make the adrenal glands start pulling their fair share of the work instead of sitting on top of the kidneys getting a free-ride. So the pituitary takes the hint and sends some adrenal hormones telling the adrenal glands to get to work and pitch in.

I understand that physiologists and anatomists are going to form a mob to surround my house and tar-and-feather me for such simplistic, dramatic claims, but for everyone else, I think you get the idea.

The pituitary gland is an incredible little bit of your brain, but it doesn't hold the monopoly on hormones. Every single one of the trillion or so nerve-cells in your skull all work by hormones. Some hormones make your nerves energetic and work faster, some hormones make the nerves kind of depressed and slow down.

Some nerves work opposite of all the others, just to be contrary and rebellious so that when they get the Exciting hormones, they rebel and do the Slow-Down. Some nerves are always such downers that when they get the downer hormones they say, "screw it, I'm gonna do SOMETHING this time!" and they get a burst of energy.

Some nerves only work with certain, specific, signals, some work from any signals. Some nerves are introverted little loners that don't have many connections with other parts of the brain but are incredibly specialized and really good at what they do. Some are obnoxiously gregarious social butterflies with millions of connections throughout the brain and are easily swayed by trendy things. They may not be right, but they sure are good at getting stuff done and getting nerves around them riled up!

Most nerves are simply in-betweeners. Their job is to send news from one place to the next. The minority of nerves are actual do-ers, sending signals to muscles or churning out some specialized hormone, like adrenaline.

Nerves and nerve pathways act like canyons and valleys and river beds, or pipes and storm drains, channeling and directing water somewhere. The more water that runs down a canyon and river bed, typically, the deeper the canyon gets, and the less likely it'll be in the future that a flood or something will change the path of the water.

Imagine the kind of change that would have to occur to get the Colorado River to change its course from the Grand Canyon! It's been carving out that path for so long and pulls water into it from such a huge area, that it'll likely take some earth-shattering event to get it to go a different way.

Even if you built some kind of dam up above Lake Powell and diverted the river through New Mexico into the Rio Grande, you'd have a hard time keeping it from falling back downhill into its habitual rut. Plus there's all the rain still falling into the canyon you'd have to figure out how to catch and divert. From a nerdy-engineering standpoint, you could build a dam all the way up to the rim of the canyon, let it silt up with sand, and then watch to see where the river and lake overflows into, and start a NEW Grand Canyon! Theoretically.

Similarly, if you take a person out of their normal lifestyle, their normal habits, their normal environment, you can divert all those pathways and habits. Lifetime criminals can become productive citizens, and addicts-from-birth can conquer what's held them down. It's not easy, but it can be done!

On the other side of deeply ingrained rut-like habits, have you ever seen a broad flat floodplain where the river just meanders wherever it wants to? It doesn't take much of a storm or flood to spill over the banks and create a new path somewhere else, and this happens regularly every now and then. Over time you get this nice fertile, flat valley floor perfect for growing things (when they're not being flooded). If we don't develop habits, or we have a life that's constantly changing, we are like that, unpredictable, though spontaneous and energetic.

Before so many modern flood control measures were built, many people living near the Mississippi River had to worry about floods. Many people still do, for good reason. Humanity was built near these kinds of rivers, like the Nile, the Tigris and Euphrates, and Indus rivers. Living near a good river where you could grow stuff meant dealing with floods.

Repeated behaviors and patterns and habits and addictions are like rivers. When you do something that your brain likes, it reinforces the nerves that made it happen. It creates a solid pathway to use again. Over years and years, this pathway dominates, and it's hard to break free of it, if you'd ever want to, especially when it's a big, system-wide, emotional thing. For athletes who perform the same kinds of repetitive tasks over and over and over again to perfection, this trait is a fabulous one.

Have you ever seen the basketball player after he throws some free-throws try to high-five teammates who aren't actually there? He's gotten into such a pattern of slapping hands with encouraging teammates after throwing a free-throw that he'll automatically reach out for them, even when they're not there. I've gotten more than a few good laughs watching internet videos of those guys, at their expense. Professional tennis athletes have ingrained into their brain all the pathways and ruts needed to fine-tune their arm, wrist, or feet, to respond to the ball coming at them in that specific way to return the ball to the other side of the court.

I used to drive a classic Mustang that my older sister and dad had purchased when my sister was in high school. Five of the nine kids in my family learned to drive a manual transmission in that car, and it was such a picky and difficult transmission that I have a lot of negative emotions associated with that stick in the middle of the floor and that eternally-long clutch pedal. We burned through just a few clutch-plates and killed the transmission at least a few times. Once you got the hang of it, it was a blast to drive. A beater, but fun to drive.

A few years after my parents sold that beat-up 65 Mustang, my younger brother bought a 1967 Mustang with the help of my parents, but this one had the typical automatic transmission. It was bright red and even had the more powerful V-8 289 cubic-inch engine in it. The 1965 Mustang had been a dull, worn out white, and had the small straight 6-cylinder 200 cubic inch engine. Despite the differences on the outside and under the hood, the interiors were a lot alike. Every time I sat down in my brother's "newer" Mustang, I automatically reached for the stick and cranked my knee up near my chest to put my left foot on the clutch. Every time. Why?

Learning to drive the older Mustang was a fairly emotional experience. It was frustrating, it was maddening, it was cool, it was a lot of fun and it was new. The original Mustang left its imprint on me and that formed many of the original pathways of driving a car. And to sit down in another car that was similar to it, activated all those old pathways every time. Did I eventually learn to control that reflex and not automatically press an invisible clutch or grab for a non-existent gear shifter? Yes, but even after I learned to control my hands and feet, I still felt the "pull" of reflex trying to do it.

Is this "pull" for a stick shift the same as an addiction? No way... sort of. But it illustrates a point. I had created this rut in my mind from a repeated action attached to some strong emotions, and it took actual effort to escape that rut, that pathway, that canyon.

One of our brain's greatest survival mechanisms is a substance called Dopamine. Another is Glutamate. Several really important parts of our mind use, make, and release these chemicals. They make you feel good. Really, really good. How much you get is supposed to be dependent on the action you just performed or the good thing that just happened to you.

If you'd lived a million years ago, running alongside your tribe after some big game, you'd feel good being with them, working as a team. When you took down the kill, you'd feel great celebrating what you'd accomplished. And then, when you take that first awesome bite of calorie-, mineral-, vitamin-rich- meat after a week of hunting, you'd feel fantastic. And you would keep doing it again and again for as long as your brain functioned. All thanks to dopamine and glutamate.

There's several other important neurotransmitters that help us feel good. A "neurotransmitter" is basically any chemical, hormone, or substance that is used by the body for one nerve to talk to other nerves. There's a teeny tiny gap in between each and every nerve, they don't actually touch. And at the ends of each and every nerve are little packets of hormones and chemicals. When a nerve is activated, it dumps those chemicals out the end where they get soaked up by the nerves closest to it.

Some nerves are just constantly going, like a little kid on caffeine, sending a continuous streaming message, until something happens upstream to slow it down and it crashes. Some nerves are always off until a strong enough signal triggers it like a stick of dynamite that finally got its fuse lit. Some are touchy and sensitive, while others are like boulders that take some earth-shattering event to get them to move.

Glutamate and other stimulants are good at forming new connections and blasting new pathways into your brain to get those sluggish ones to move, and Dopamine is the paving reinforcer that helps solidify the pathway into sticking around.

Throughout our body, away from the brain, we have nicotinic and muscarinic type nerve receptors. They're named after plant chemicals that happen to activate those nerves, nicotine and muscarine. Nicotine, as you probably learned in elementary school No-Smoking messages, just-so-happens to activate one of these types of receptors, so some scientist somewhere got really creative and called them nicotinic receptors.

These receptors cause large parts of you to relax, or cause a lot of the muscles and organs in your body to relax, as if your brain were telling them to. It gives an externally-caused sense of peace and well-being. This plant sends a message, rather than waiting for your brain to do it. "Chill out guys, everything's cool right now, just smoke some more of this good stuff." Nicotine hijacks those receptors without your brain's input, contributing to a false sense of security and peace.

Ever notice how people go for a smoke when they're nervous or anxious? That would make sense. There are some nerves that get jittery with nicotine, which is weird, but that's just part of the cocktail.

Your brain still gets the feedback from your body, "Hey we're all cool down here, I don't know what you're so stressed out about." If your body is relaxed, your brain decides, you MUST be relaxed. So, Brain thinks to itself, "Huh, oh well, whatever you're doing, it's working and keeping us safe, so keep doing your thing. Oh, and let's do MORE of that." Except, the true problem or stressor isn't ever resolved. So you hide it with smoking, or sex, or booze, or anger, or TV, or...something.

Thus an addictive habit is born.

Muscarinic receptors don't really cause much of an addiction problem. They're named after muscarine, an ingredient in poisonous mushrooms that causes convulsions and death, so I'm not gonna waste much time talking about them. Our body obviously makes something like muscarine, but the power of our body's own muscarine is much, much, much weaker than the mushroom version. Like a chihuahua's bite versus a pit bull's. One lets go, the other doesn't.

There's actually several types of nicotinic and muscarinic chemicals and receptors throughout the body, and other classes and varieties of nerves, and we've figured out how to use these receptors and the chemicals that activate them for various useful drugs and pharmaceuticals. It's pretty amazing and cool what they can do, but you can read Wikipedia for more on that. Go look up Curare. Cool stuff.

Again, to all the scientists reading this, these are very simplified generalizations, I realize that. Going too deep into biochemistry and neuroanatomy isn't going to help people read this book and I'm risking losing my audience as it is, so I'll spare you ALL the technical jargon. I'll cover some of the major chemicals, the ones you'll hear about on the news from time to time, see on the shelf in your health food store, or while talking to your doctor. I'll especially focus on those involved in addictive behaviors and leave it at that.

Amongst the weaponry of the brain are such diverse elements as GABA, Glutamate, all of the Opioids, Catecholamines, Monoamines like adrenaline or epinephrine, Substance P and its Nociceptors, the muscarinic and nicotinic buddies, various other amino acids and minerals and more. It's a soup in there, all fighting to influence and tell your brain what is going on in the outside world. Not only that, the typical five senses, the position of your body, even the position and control of your muscles have an influence on the brain.

Let's buzz through a couple of these, a crash course if you will.

Glutamate - is the most prevalent and common Stimulant. It's a simple amino acid, a normal part of the protein in your food, and you have taste buds that can pick it up to tell you if you're eating protein. There are some reports that eating too much glutamate too fast can make many people feel shaky or give them headaches. If you add some salty sodium to it, you have....MSG, monosodium glutamate. It's tasty for a reason! Stimulants are like little explosions. They blast new pathways into your brain. But they're not very specific. Engineers blast specific little explosions when they are digging a tunnel or carving a freeway out of a mountain-side, but glutamate just kinda goes off wherever it can. Sometimes the new pathways are helpful, but when we get too big of doses of glutamate, these new pathways are not always helpful.

GABA - or gamma-Aminobutyric Acid, is the most prevalent inhibitory, or depressive transmitter. It slows stuff down. It slows your brain down.

Dopamine - we addressed this guy already. He's cool but gets a little carried away sometimes. He's the transmitter that tells you when you're doing things right, but he has a tendency towards flattery and brown-nosing that gets out of hand and gets you into trouble. And the more you listen to him, the louder he gets. (yes, dopamine is a guy, because those kinds of guys always have to be right and dopamine guys are lame like that. That or it's because I'm a guy and I'm gender-centric, you can pick).

Opioids/endorphins - There's a good handful of natural opioids made inside your brain and body. Opioids are named after a plant ingredient that acts very similar that triggers the same nerves: Opium. Lots of people have heard of endorphins. Endorphins are the naturally occurring opioids that your brain uses to help you feel better when something is stressing you out. It gives you a runner's high. It suppresses pain to get you out of danger. And it's really powerful. Humans have been using external opioids from poppy seeds for hundreds of years, and more recently we've come up with some incredibly potent and cheap-to-make copycats.

Opioids typically include morphine, heroin, opium, oxycodone (Percocet, Oxycontin), hydrocodone (Vicodin, Lortab, Norco), Fentanyl, and more. These are very addictive and very potent because they take away pain. Not just injury pain, but the normal day-to-day average emotional pain we all have to deal with. If there were such a thing as a magic place where everything was good and wonderful, opioids might be the closest we could get to it. In reality, they don't always block pain but they can make your brain not care much about it. Same difference, right? Not really.

Epinephrines - These are your adrenalines and yes, there are more than one kind, acting together in different parts of your body. Simply put, they kick on your basic survival instincts and shut down everything else. Epinephrine "receptors," or triggers, do different things in different parts of your body. In your muscles and brain and eyes, epinephrine floods them with blood so they can work really well, really fast. In your digestive system and immune system, epinephrine shuts down muscles and blood flow, slowing down digestion and the repair work your body is trying to take care of. "You'll have time to eat after we don't die!" it says. Super-human strength to lift cars off of trapped babies? Thank you Epinephrine.

Epinephrine kicks our inner "reptilian" brain into high gear. We don't do our best thinking when we're drunk on adrenaline. We're not usually our nicest either. We become scared, fearful, cranky, angry, hostile, aggressive, and selfish. And it doesn't take much to give us a shot of adrenaline. A sideways glance from a co-worker. A whisper and a finger pointed at your direction. A crowd's attention on you. Anything that's a potential threat will drop some adrenaline in our brain. Even a PERCEIVED threat, not a REAL threat, will do the same. We just have to THINK we're under some kind of attack, no matter how small it really is.

If you're constantly stressed, that could be because of your environment, or it could be because of how you're reacting to your environment. By "environment," I mean your job, your family, or your life in general. Not the forest, and mountains and trees kind of environment, but everything going on around you. If you're constantly stressed, your immune system never gets much of a chance to heal your body and clean out the garbage. Or your intestines don't absorb food the way they're supposed to. You start craving more sugary-sweet stuff for immediate fuel, and less of the vitamin-rich healthy stuff. After years of this, you're pretty sick. Diabetes, heart disease, auto-immune problems are all related to chronic stress and anxiety. Dang you, Epinephrine and your Henchmen! And thank you, too, y'know, for keeping us alive.

Cocaine is a powerful drug because it triggers all the nerves in your brain to just dump their epinephrine out into the rest of your brain, uncontrolled. You get a crazy rush. Your eyes will dilate, your heart will start beating fast, you may get crazy aggressive or really horny, or both. You will definitely not be thinking straight. And the rest of your brain will start desensitizing those nerves and cells so that it doesn't happen again, so you'll need more cocaine the next time to get the same high.

Serotonin - This one is important for your overall mood, sleep, memory formation, appetite, and lots of other complicated things, including social rank and dominance. Weird that the amount of a chemical in your body that's used for sleep has something to do with social dominance. Is that weird to anyone else? Maybe the relaxation and brain function it promotes has to do with confidence, and confidence aids our ability to interact comfortably with others. That sounds believable, but I'll be honest, I just pulled that out of thin air.

It's synthesized from regular amino acids found in regular proteins, not just the Thanksgiving turkey that's putting you into that coma. Lots of sleep medications mess with serotonin and can be habit-forming. Your brain becomes dependent on the extra meds to calm you down and isn't satisfied anymore with the natural amounts your body would normally make.

Your outer brain doesn't learn what it needs to learn to shut down your inner survival brain. Those pathways that slow down your stress-brain either get weak or are lost when you depend on sleep meds that you don't truly need. But they do work GREAT for those who really need it. Because sleep is REALLY important. That's when your body finally has a chance to stop being stressed and repair everything it encountered during the day.

Oxytocin - The "Love Hormone". This is important in all aspects of childbirth and nursing. It helps women get "in the mood" and helps men stay monogamous. It's true! Give married men a shot of oxytocin and they will keep more distance between themselves and other women. More so than single men given the same shot! And, the brains of men being given oxytocin will react differently to being shown pictures of their partners compared to pictures of strangers or even female acquaintances, as in, they'll feel more affection and attraction when being shown pictures of their partner, vs other attractive people, as long as their partner is a source of steady oxytocin.

It's important for bonding and feelings of love. It helps people feel connected, even as friends. When you are enjoying social interaction with other people, you are enjoying oxytocin. It's critical to cooperation and being a member of society. Your body makes more of it during skin-to-skin contact and can have many of the same effects as alcohol when it comes to overcoming inhibitions and working up the courage to do things you wouldn't normally do in front of other people.

Children are filled with oxytocin when snuggling with their mother. Adults (and even children) usually get a shot of oxytocin from their brain when they see a woman's breasts or the reproductive organs of either gender. Women will get a rush of oxytocin when they expose their breasts or large amounts of their body. So will men. Why would people risk a charge of "Indecent Exposure," flashing their full bodies to strangers if they didn't get some chemical hit out of it? These chemical hits, whatever their source, also leave imprints on us that can last for years, or even a lifetime.

As much as many people wish we could all "just get over" the fact that all women have breasts, or that nudity is normal, and that there's nothing special about this, biology is always going to pull at us in ways we may or may not like when we see them. Clothing has been a part of humanity since long before history was recorded and since long before any semblance of civilization was formed. Nudity became "unnatural" long ago, and remains a potent cue for triggering our sexual and reproductive instincts, through oxytocin.

I've known more than a few men who were determined to avoid female pornography at almost all costs, but still needed some sort of outlet or release for the hormones building up inside them, so they turn to male pornography or various forms of masturbation and cross-dressing to still get their needs met.

It's easy for a stressed and anxious mind, dealing with an overflow of hormones and a lack of oxytocin in their life, to find some way to substitute it. They think to themselves, "It's not REAL porn if I'm looking at people like me, my own gender." Or, "It's not bad if I'm ogling myself, no one is getting hurt." This is an easy lie we tell ourselves. We are still very much affected by these levels of intimacy and visual cues, and they leave lasting impressions on us, that then affect how we see the world and people around us.

Every relationship in life requires an appropriate level of physical contact. For some, it is limited to a handshake or cheek-kiss when you greet a friend.

For business partners, a handshake is usually expected. For family members, a hug is an expected part of the relationship. For a pair of people that are beginning to get to know each other personally or intimately, a kiss and holding hands, or long hug is needed to reinforce that relationship. For intimate partners committed to each other, to building a family or to having children, emotionally connecting sexual physical contact is expected and needed to keep the relationship functional.

To have a mismatch of physical contact for the relationship, your brain gets mixed signals and becomes stressed. The oxytocin isn't matching with everything else and the relationship will become strained, or our other relationships can be strained. For many relationships, NO physical contact is the expected amount. Oxytocin is the hormone that binds people together, or keeps them appropriately apart.

How many people, not getting the level of affection they think they need from their partner, will turn to their children, or strangers to get it? Even if it's not sexual, it is dangerous to let your heart become too committed to another person. Mothers, eager for male companionship but trying to avoid the responsibility of fulfilling their husband's sexual needs, will sometimes substitute their sons for such a companion, placing pressures and expectations on these sons that aren't healthy for anyone. Father's and husbands will often do the opposite, looking for someone who can fulfill their sexual needs without having to be responsible for the other person's emotional needs.

Stirring up extra oxytocin through pictures or fantasies, or being in a relationship JUST for the physical affection —just to get the good feelings from it— can be a bad thing for us emotionally, because we'll get used to higher amounts of it than our body would naturally make, and we can get hung out to dry trying to keep those high amounts of it still flowing without all the other reinforcement needed for a stable relationship. Friends with benefits, or hookups, will eventually hurt our brain in serious ways. That mismatch in the opposite direction will cause stress to our brain and mess us up.

Just like using drugs or drinking at a young age can get kids and teenagers more seriously addicted, getting even mild sexual affection at too young an age can get kids and teenagers addicted to being in relationships before they're smart or mature enough to stay out of trouble, or keep them in relationships that are bad for them. They never learn how to function as an individual.

Some children, who never receive enough physical affection from their parents won't learn how important it is to forming relationships and may struggle forming relationships throughout life until they can learn what they need to as adults.

Yes, parents will screw up. Every parent will. If you go through life struggling because your parents were imperfect, or even just plain bad, join the club! You belong to the inclusive group known as Humanity! It's important to learn what you missed and compensate for it, but it doesn't do much good holding onto resentment and anger for your parents.

Back to Oxytocin.

The drug that hospitals give women to induce labor, or speed it up, is called Pitocin and is just man-made oxytocin. You can get more of it from a nurse or doctor than your body would make on its own.

Losing a partner, or source of physical affection and bonding, causes us to actually go through physical withdrawals when our source of oxytocin dries up. Depression is common immediately after a breakup or death of a loved one. There's no more oxytocin telling us we're doing something good. That withdrawal is a big reason why so many people rebound into a new bad relationship right after the ending of a previous long-term relationship.

How many of us have done that or seen friends or family do that? Yup, just about all of us.

That withdrawal and the accompanying depression probably explains many of the stories we hear about elderly couples who've been married for many, many....many years. It's not uncommon to hear about one of them passing away and very shortly after, the other spouse passing away, too.

Oxytocin is like Liquid Courage, but ultra-self-made. Do we sometimes act in very foolish and silly ways when we're newly in love? Oxytocin. Do we do crazy things to impress the person we have a crush on? Oxytocin.

It's powerful, and also easily thrown off and out of balance by alcohol.

Alcohol - Our "friend" here does a whole bunch of things to mess with all the other neurotransmitters. It suppresses glutamate, and generally turns brain activity down a couple notches. It promotes GABA. It enhances and magnifies the effects of dopamine and the opioids. It messes with serotonin and just about everything else. Alcohol isn't synthesized in significant amounts inside the body and is usually acquired through fermented, carbohydrate-rich foods that also tend to be calorie- and vitamin-rich. Alcoholic drinks were a valuable and useful source of calories, vitamins, and clean, germ-free fluids for ancient peoples.

If you get a chance, check out the story of the North Pond Hermit who lived alone in the woods of Maine for 27 years and would survive the winters by fattening up on stolen cheap alcohol. But that alcohol, it does some pretty awful stuff to our liver, brain, and other organs. There's some true good use to it, but today we tend to abuse it more than anything. It is, indeed, the most commonly abused substance on the planet and has been for, I don't know, a few thousand years. It's a pretty natural process to make it and there are some benefits.

Grains and fruits—the stuff that we live off of—can ferment from the germs and yeasts that live pretty much everywhere, even our skin. Fermented juices and foods are usually high in calories. They're often full of vitamins and antioxidants, and the whole fermentation thing can pretty much happen to food without much human intervention (although it may not taste as good without our help).

In modern times and societies, we don't NEED alcoholic drinks for the same reasons we used to, but it does help explain how alcohol became such a prevalent part of humanity. Because it is so good at turning down our senses, thoughts, morals, and the stresses of life that we should be dealing with, it becomes a very addictive substance. It also has a strong ability to ramp up our drives, like our drive for sex, by turning down our higher thinking and letting our primitive inner brain do whatever it wants.

Our inhibitions disappear and we become a different version of ourselves, not a truer version, or a more real version, just a different version. Who we are when we are sober is valuable and more important than who we are when we are drunk, and most people recognize that. It's when we are drunk more than we are sober, when we can't function socially without getting buzzed, or when we make life-altering decisions while under the influence that it becomes a serious, serious problem.

Some new research has shown that a handful of drinks per day can severely hamper brain growth in juveniles.

Alcohol is super popular, I won't spend much time condemning it. It is good to know what it does and how it affects us.

Sex-hormones - A number of other hormones have significant impacts on our thoughts. Testosterone and the related androgens, estrogens, and progesterones are the key players. Men and women have both in various mixes and amounts and they are influenced by environmental and genetic factors that are as diverse as our family life, our social circle, the food we eat, the environment, our past memories, our work circumstances, and even our conscious behavior.

Whether we live with—or grew up with—other men or women will alter our sex hormones. The movies we watch, the books we read, the foods we eat all affect our sex-hormones. What we look at, what we think about, what we are exposed to can all stir up sex hormones. Smells, memories, foods, settings, textures, and so many other things can stir up sex hormones and ramp up a person's sex drive. Talking about sex, even in a matter-of-fact, un-emotional way can still stir up a person's sex drive.

Have you ever noticed in a group of people, whether it's with friends, co-workers, or strangers, how once the topic turns to something sexual, the conversation stays sexual? It usually requires someone consciously changing the subject to get it off of the topic.

Once I went on a hike with some friends and their kids. My good friend is responsible, hard-working, loving, devoted, and always appropriate. We encountered another set of hikers and it was incredibly obvious that one of the women was not wearing much under her shirt. No one could miss the fact. No one in our group said anything about it, we just gave each other some looks that expressed what we were feeling: "Whoa, did you see that?!"

Despite the shortness of the experience, there was a subtle shift in my friend's way of talking to his wife. It was much more teasing-like, with flirting, and the kinds of comments a typical husband makes when trying to arouse his wife subtly in public. Did he realize what had happened? I'm guessing not. It was probably unconscious and subtle enough that he probably didn't even notice his behavior changed. I noticed, as an outside observer. We can learn to watch for this in ourselves if we are careful and compensate if we need to.

Since these hormones oversee what is probably the most dominating drive to all living things, we need to beware of them and respect them, carefully. They can and often do take over our lives, and while sex isn't a "requirement" for survival like food and water, your body was designed primarily to pass on your DNA and will do whatever is in its power to satisfy that urge. Resisting that urge is often even described as "painful", "torture", or "miserable." If that's what it feels like to resist our sex-drive, why and how would anyone do it? I'll explain later in this book.

We can achieve mastery over it, but it takes a great deal more insight, practice, and knowledge than many of us will achieve. For the large majority of us, it is better to build our lives in a way that allows us to satisfy our drives in appropriate, constructive ways and turn the volume down on the inappropriate things that stir up the drive, rather than try to avoid it altogether.

Substance P - this is a chemical that does one thing: hurts. It tells your brain, "Ow." The substance doesn't actually hurt, it just turns on special nerves called nociceptors, that have the sole message of telling your brain, "Physical damage! Alert!" We can train our brain to not over-react when it gets these messages. Our brain can also get tired of the alert and just turn down the volume of that nerve by desensitizing it. We can even train our brain to enjoy pain, like a stimulant, like a drug.

Not all Pain comes from Substance P or Nociceptors. Sometimes what we perceive as "pain" is more accurately, just over-stimulation of one of our senses. Like turning a light on suddenly in the night. It "hurts" our eyes, though there's not always "pain" or damage actually occurring.

Spicy peppers work their wonderful magic by sticking to the nociceptors in our tongue and activating temperature-sensors. This makes your brain pay some extra special attention to what the heck is going on down in your mouth, and thus it pays more attention to the taste of your food, and the food tastes mmmm mmm good. Until you add too much chili powder and are suffering for the next 10 minutes. Flaming Cheetos?

There's a reason why they are so popular: Your brain is forced to pay attention to how salty, sugary, cheesy they are and then that nice kick of glutamate (MSG) adds an extra punch! Oh man, I love Flaming Cheetos. Before every test in college, I'd buy myself a bag of Flaming Cheetos and a chocolate milk. I swear it helped, but I can't say I recommend it to all of you, because: junk food. But still, mmmmm.

When we work out, it causes physical damage to our muscles, and we get substance P floating through our Burpee-Crossfitted biceps. If we can learn to appreciate and enjoy that feeling, we can learn to overcome that pain and build muscle and become stronger simply for personal desire. When we become stronger, we are rewarded with the ability to do more things (and be more attractive and earn more oxytocin and dopamine!) Did you know that exercise can be addictive? There are a few good reasons for it.

Pain can go too far, as well. Some of us, for a whole list of different reasons, can become addicted to pain, and we'll actually cause damage to our body on purpose, just for the stimulation and nothing else. It's one thing to enjoy cuddling with a partner even if it's uncomfortable. It's another thing to be uncomfortable on purpose, without the cuddling, just because.

Or what if we are dealing with enough negative emotions making our hearts "ache" and we can't stand it, but huh, when I scratch and cut at myself, the skin-pain takes my mind off of the emotional pain and it feels better than dealing with the stress that I can't resolve. I've had more than enough friends who "cut" and it can take them some time and help to figure out why they do. Sometimes we use some sort of "controllable" pain, or self-inflicted pain, to get our mind off of the deeper emotional pain we don't know how to resolve.

Pain, if it happens long enough, can get stuck in an "on" position. Like a canyon or rut I mentioned before, or like a corroded water valve, it doesn't turn off once it's been on for too long. For example, the other day, my kitchen faucet suddenly turned on by itself when a seal cracked and broke. I had to have a friend bring over a large plumbing wrench and carefully turn off the water to the faucet under the sink because the knobs and valves had been stuck in the "on" position for so long. This can happen to nerves. Doctors call this "chronic pain." When you've been dealing with back pain for so long that it's just become a part of your life, sometimes that pain will never turn off even when your back gets fixed.

For those of us who don't have to deal with chronic pain, imagine how hard it would be to be nice and pleasant to others, when your body is just constantly hurting, no matter what. Could you be nice dealing with that? I get severe headaches if I'm not careful with my posture or sleeping position, and when I have a migraine, I'm not much fun to be around, even though I try my best to be nice. It stinks. And some people have to be nice when they are feeling awful. Some people just can't keep the niceness going and they don't know how to make the pain stop. It's hard!

Pain is still somewhat mysterious and we can't always figure out how to stop it or even what's causing it. For those of us who have to deal with chronic pain, there are lots of things to try. Talk to your doctors and specialists. Most are fine prescribing narcotics and opioid pain meds because they work so well, but keep your mind open to trying new things, and don't give up hope and don't stop learning as much as you can. Many people have found surprising and healthy ways to get rid of their pain. Sometimes it's by changing your life to get the rest of your body stronger, or changing your diet, or even some ingrained habits and simple behaviors.

As we learn more about trigger points, we are finding that many issues we thought were related to joints, tendons, vertebrae, or other bone problems are actually due to over-active muscle fibers that just don't want to relax because for years we constantly kept it tense until we don't know how to relax it ourselves.

Behavior - How is behavior a neurotransmitter? It's not, but it fits in with the others. It's long been known that when we ACT happy, we tend to BECOME happier. When we ACT aggressive, we tend to BECOME aggressive. There's such an intimate connection between brain and body that they influence each other. It's not just a one-way flow of information. If I spin a fan, it blows air. But, if I blow air into a fan, it spins! I can use electricity to spin a motor, or I can spin the motor to get electricity. We can use our brain to control our body, or use our body to control our brain, to a degree. Gritting my teeth and tightening my shoulder muscles is as likely to make me stressed and flood my body with stress-chemicals as an actual injury to the body.

There should be no discounting the effect of behavior on our mind. HOW we act has such a strong influence on our mind.

Exercise, physical activity, and just plain USING your body to do something that gets your brain working can be an amazing escape from our problems, or as a way to churn our brain into solving them. Whether it's running, working out at the gym, some handicraft or artistic creation, a walk through nature, or solving a puzzle, if you get your physical body involved, your mental body will work better.

Our Senses - What we touch, smell, feel, see, and hear all have genetic and emotional effects that will affect our hormones, our thoughts, and our emotions. This probably won't surprise most people, but our environment has an effect on us and for good reason (Thank you Captain Obvious!).

When we are in situations that we can survive and even thrive in, our body will take advantage of it and enjoy it. When we are in a situation that we don't think we can survive in, our body will push us to either change ourselves or change the environment. Some of us feel like changing the environment is easier while some of us feel like changing ourselves is. Whichever one we get success with early on in life is usually the one that dominates for the rest of our life, though it doesn't have to. Sometimes we'll try to change what's around us instead of changing ourselves. Sometimes it's the opposite.

There have been numerous studies showing that when we are exposed to negativity, incivility, violence, or any number of other stressors, even in gentle laboratory settings, it affects our mind, our hormones, and our emotions, for hours or even days afterward. When given a scrambled set of letters, people are more likely to unscramble those letters into words like "murder" or "death" if they had just looked over a list with a sample of neutral and negative words in it just before, compared to people exposed to a list with purely neutral, or positive words. Negativity is contagious. Fortunately, so is positivity.

Hardly anyone likes going to the dentist to get their teeth fixed, most usually for the scary part of getting numb or the high-pitched sound of the drill (personally, I think it sounds awesome, like a mini jet-turbine in the palm of my hand!). Oddly enough, dentists are nervous when their patients are nervous. When it gets measured, we've learned that dentists will experience an increase in heart rate, blood pressure, and overall tension while they are giving the anesthetic! Turns out, we don't like getting people numb!

The shows we watch, the music we listen to, and the articles we read have a lasting effect on us, whether we want them to or not. Very few kids who play violent video games will become school shooters, but those violent games do have a negative effect on our mind in other ways. Watching political commentary shows tend to make us stressed and angry with their spin on things. Scary movies leave our adrenaline pumping for a few hours or days, and we'll often respond in less than ideal ways when it counts, while we are under its influence.

Many studies have shown that the music you listen to can speed your heart rate and increase your blood pressure, or it can slow your heart down, and decrease your tension. Music has been a part of every culture of the world and has been used in almost all of the world's cultures for therapy and healing of various kinds.

When we are around those who are scared or stressed, it stresses us out. We may pick up on it from non-verbal cues, or body language, or we may even pick up on it from our noses.

Most biologists know about animal and insect pheromones, those chemicals that a creature gives off to communicate with other animals and creatures. It's how ants can talk to each other, it's how many mammals know when it's mating season, and it's even how some plants in a forest can warn each other of threats. Human pheromones have been mysterious unicorns, for the most part, but an interesting study at a dental school in 2018 gets us closer to figuring them out.

The researchers took donated t-shirts from a group of students and had them wear them during a stressful test and another during a boring lecture. The shirts were then sprayed to hide and body odor, and placed onto mannequins the students were using for practice. Students who worked on the "stressed" mannequins did just a bit worse in their practice than the students who worked with the "relaxed" mannequins. It's some hint that humans, literally, can smell Fear, and it tends to make us perform worse.

Summary:

Just about everything we meet has an effect on our body and our brain, and thus has an effect on our thoughts. Some of it is controllable, a lot of it is not—at least not directly. Whichever methods we first find success with to deal with stress and negative emotions early on are usually the methods we stick with for years and years, whether they're good, constructive, healing or not.

4

Who We Are: A Split Between Our Brain, Our Personality, and Our Decisions

We've talked about how chemical and hormonal substances can alter our thinking and play a major role in our brain's functioning, let's now talk about the levels of that functioning.

As Homo Sapiens, we have the most developed and most organized brain of any life form on the planet. By itself, a human mind is the most complicated and complex clump of atoms in the known universe.

As life has evolved and grown on this planet, chemical reactions inside of single-celled organisms were enough to ensure survival and reproduction. A bacterium bumps into food, the proteins on its surface stick to it and fold in such a way that food gets wrapped up inside of the bacterium and chopped up by other proteins and enzymes. Every time. I put enough heat into something, it ignites into fire. I drop an apple, it falls. A dangerous peroxide molecule bumps into a peroxidase enzyme and it's chopped into harmless water and oxygen—lightning fast—like a mouse-trap going off a thousand times a second. Pure and simple laws of physics and chemistry.

As life forms started working together with other life forms, they started forming colonies, and then certain cells started specializing in certain kinds of work based on their environment or how close they were to the environment. Cells on the outside would become watertight, while cells on the inside got better at digesting food and sharing. When things got too complicated and too far apart for communicating by chemicals and hormones, some cells got good at changing shape or sending electric pulses. Some of these became nerves and are found in everything from insects to apes.

Clumps of nerves were eventually needed to organize the chaos. The needs of the body determined the needs of the brain, and then the success of the brain determined the success of the species. Honey bees have a pretty small cluster of nerve cells that take care of their little bodies but are complicated enough to allow cooperation within a hive society and even make for odd useful behaviors like how to "dance and wiggle" to point directions to food.

The size of a brain doesn't have to do with being smart, nor does it translate directly into intelligence, it just does what it needs to do and takes up the size it can or needs. For the large size of its head, a tiger's brain isn't all that big, and for the relative size of a whale, its brain is teeny compared to its body, though their brains are still bigger than ours.

As animal brains developed, there were two general regions of them: A cerebellum and spine which fine-tuned most of the muscle control and reflexes, and a "reptilian" Midbrain which received all the input from our senses and environment. It's all pretty reflexive with little actual "thinking" going on. As animals became more advanced and complicated we developed a super-system on top of our reptilian mid-brain. Now we have a "Go!" system and a "Know" system. The Go system is the "Fight or Flight" system. Our senses predict danger, a flood of hormones surge out, and muscles start springing to action to either dominate or run away. Our Know system is the "Rest and Digest" system. Occasionally, we're overwhelmed or confused and neither one works and we freeze—like a deer staring at headlights—and we get run down by whatever it is that's coming at us.

If the body reports that food is nearby, the brain says to go get it. When everything's going just peachy, when all is safe and calm, and there's food enough and perhaps to spare, then the brain says, "Let's get busy passing on our genes, let's go find a mate!" Very reflexive, very instinctive. Very, "animal." Your body does what your brain tells it, and your brain is mostly the result of millions of years of animalistic evolution and instinct.

Even language, love, nurturing, social hierarchy, use of tools and technology, and forming prejudices from infancy are things we share with animals.

These two systems, officially called the "Sympathetic" and "Parasympathetic" nervous systems, are in a constant balance of sorts. One is the gas and the other is the brake. They keep us going at the speed that best serves our needs, and they decide about or at least 95% of our decisions, without our input, without conscious choice. These are our instincts, our reflexes, and drives. I made that percentage up, but the point is, that the HUGE majority of our decisions are made reflexively by our brain, driven by environment and biology. (Did you know that 87.6% of statistics are made up on the spot? I made that statistic up right there, but I betcha it's true)

What's the difference between instinct and reflex? When I tap that soft spot near your knee and your leg kicks out, that's a reflex. The signal goes only so far as your spine before a message goes back to the knee to jerk, and only after it's all over does the message get to your brain that something even happened. When your finger touches something hot and you jerk away before you even realize what's going on, that's a reflex. Almost.... almost, out of your control.

Instinct is more complicated. A baby human is born and starts searching for something to suck on, that's instinct. A mother turkey nurtures and protects ANYTHING that makes the baby-turkey-cooing sound, that's instinct. Baby turtles hatch, dig up through the sand, and head towards the ocean. Instinct. No one taught them anything. It's just....there.

An instinct is a REALLY complicated reflex. Where that boundary is between "reflex" and "instinct" is anybody's guess. There's a healthy debate among scholars of all kinds about it, and there will be for a long time yet.

We have a very instinctive brain, much more instinctive than I think anyone wants to really believe. But, primates and certain other animals have developed a part of our brain that goes an extra level above instinct, from just a REALLY complicated reflex into a COSMICALLY COMPLICATED reflex. It is so complicated that, most times, we can't see the cause that created the effect.

A customer at work snipes at us for an hour and life goes on. Yet later, when we're alone and life is calm, we'll be on a Ben & Jerry's binge and feeling bad about ourselves for being so gross that we ate a whole carton in one sitting—even though it helps us feel better from that dumb encounter still lingering in our memories somewhere behind the scenes, kept alive by the original flood of stress hormones that's still dissipating from earlier that day. Or perhaps it was something that happened yesterday or last week! Yet, you won't remember the incident or know that THAT is what sparked it.

At what point do we decide that we have "Free Will" over our body and its Instincts? Who knows? I've talked to neuroanatomists, neuroscientists, neurosurgeons, psychiatrists, philosophers, and psychologists and no one can give a definitive answer. Certain religions promote that we do indeed have free will, and even many scientists will contest that we do. How much Free Will do we have? Most of us agree we have at least some, but scientifically, we're not quite sure. How much do our genes and our environment control us?

How much control does our "spirit" —or whatever it is that makes the final decision— have on our body? I would claim that it's barely there, if, you believe me that it's there at all. It would be such a small influence that we cannot yet see it or accurately measure it, and that level of control is EASILY lost to our animalistic minds and our animalistic instincts, reflexes, and addictive behaviors. It's pushed forward by hormones that reward us for anything that feels good at the moment.

Many people have looked for that part of our brain, that tiny pin-point part that is free from external stimuli and can make free-choices. We haven't found it yet. I doubt that there's a single spot, but more of the whole outer cortex, the outer layer of our brain that all contributes. Our Moral and Rational brain. The part we SEEM to have some control over. This is the "Know" system.

We can take control of it. With the right amount of knowledge, teaching, some outside help, and years of bumpy, rough, practice, we can learn to control our brain. We can learn that we can create happiness. Genuine happiness through conscious thought and a little bit of cleverness.

When' I'm feeling sad, if I smile, one part of my brain controls the muscles into a nice curling of the lip. Another part of the brain picks up on this. It's like, "Oh wait! We're smiling! Why are we smiling, can anyone tell me? No one? Well, we must be smiling for a reason, alert the Positivity Nerves, we must be smiling for a reason, so let's figure out why we're smiling!"

The "Positivity Nerves" (no there's not really a set of nerves called that, this is just to give you an idea), who fell asleep on the job, are jolted awake and have to come up with reasons why the body is smiling. "Must be a Tuesday, sir! Oh, and the sky is blue, that's always a reason to smile!...and we're not....dying! Yup, that's worth smiling over." And the overall effect of all the hullabaloo is that it works!

Those reasons for being happy are just as legitimate and valid reasons for smiling as any other. When researchers scanned the brains of people "pretending" to be happy with "made-up" reasons, versus others who are feeling genuinely elated, there was no difference! Happiness is happiness. Optimists aren't blind sheep ignoring the real world, they are just choosing to focus their energy on the parts of their world that are good, while pessimists are focusing their energy on what is wrong or missing. Both are correct in their observations, and with different results.

Only imagining that you are eating ice cream activates many of the same parts of your brain as ACTUALLY EATING ice cream. Part of your brain is imagining the ice cream and sends "ice cream" signals to other parts of the brain and body. Those parts that just got the message can't tell much of a difference between a real message and a faked message, so they respond as normal and treat you as if you are eating ice cream. Like a reflex. This can be a good or bad thing, depending on whether the reward is REAL or not.

"Not dying" is a fantastic reason to be happy. Exercising is a great reason to be happy. Building things up and creating something new is designed to make you happy. Creation, though a long and often painful process, ultimately leaves us feeling proud and overjoyed while destroying can leave us feeling stimulated in the short term, and depressed over the long-term.

Taking a hit of some drug, or behaving in a way that shuts out the droning noise of the world around us, can also make us feel happy, but there's a time limit to how long you can do that, before you've run your body ragged and sucked your brain dry and are having to deal with a broken body, broken relationships, and a broken life.

5

Habit and Rituals: More Than Just Autopilot

When you've got some sort of habit that you repeat over and over and over, and it has a generally or immediately positive reaction on your brain, that trench of thought gets reinforced and built up and a reflex is created. If you normally step outside the back-door at work to have a smoke, then every time you go out the back door, whether it's to have a smoke or not, your brain gets put on alert. "Back-door time folks! Get ready for a cigarette! Nerves, organs, you know what's coming, let's not waste time waiting for the actual cigarette, just get going, you know what to do!"

When no cigarette actually arrives, you feel something missing. Some part of the picture is not whole. It's like "home" but not quite the same. Like a parallel universe where some things are similar and others not. It's unsettling. You think to yourself, "I'm here, everything else is here, I just need a cigarette and life will be complete!"

Add the actual substance reactions your body is used to enjoying to this familiarity and voila, an addiction is recognized and you have an uphill battle ahead of you trying to get control of this brain of yours.

If you have a cigarette with your morning coffee, then every time you sit down with coffee, whether it's to have an actual cigarette or not, your muscles, your organs, and your thoughts are automatically entrenched in the ritual of pulling out and lighting up a cigarette. Just like my left foot would reach for the imaginary clutch in a classic Mustang. Habit, ritual.

This guy named Pavlov did the same thing to some dogs he was studying. He would ring a bell right before he fed them. He noticed after a while, that just ringing the bell would get the dogs salivating and drooling.

For as stupid as we think fish are, the ten or so Neon Tetras in the fish tank right next to my front door, will all gather and cluster together near the top front of the tank when I walk in the door home from work, and stay there until I feed them, because that was my routine. I walk in the door and feed them. Now they expect it.

If someone is used to sitting down at the computer in their room and wasting time on internet porn (or video games) for the next 2-3 hours before they HAVE to go to bed, they are going to sit down, click click click, and they're lost to the world for a time. Ahhhh, that picture of someone showing their most private and intimate parts gives you a nice shot of oxytocin and dopamine together. Human love and connection in its most falsified form, but it's easier than the real thing by a million times and —in the short term— safer and less risky to the emotions. And thus, in time, you get sucked in every night just by walking into the room.

And these habits and rituals can get really complex! Some drug addicts around us have figured out just the right words to use at the dentist's office or doctor's office, to convince the doctor that they're in some sort of pain, and they likely have some visible indicator of it: a broken tooth, a chronic limp, which doesn't actually cause pain, but it's a great way to convince someone else that they need pain meds, like hydrocodone, oxycodone, or more. They develop a ritual, or act, that they can repeat to each new doctor to get what they need. A ritual they've practiced and refined over a few weeks or months.

A few years ago, I got alerted by the state Controlled Substance Database, that an adult patient who'd come into my office and convinced me that she had actually pulled out her own loose tooth —and maybe she had!— had been getting narcotics from somewhere between 50 and 100 different medical and dental offices over the past 2 years.

4 pages of narcotic prescriptions were in her history. You could track her habits. When she received a bigger prescription she could hold off for a month. A smaller prescription would only last her 2 weeks. A few doctors gave her so few tablets she was at another office that day or the next. Imagine how much time she spent searching for offices and actually going to them to get her "fix," and how complicated her routine, her ritual, had become to search for and find unsuspecting offices.

This ritualizing is an important part of human behavior and a key part of addictive behavior. Like a complicated reflex, easily disguised as instinct, a ritual is incredibly complicated and diverse. It is the pattern, or set of patterns that your brain draws on to get you ready to do...anything!

Whether it's sitting down in your car and reaching for the gear shifter, or getting ready for bed, or getting ready for work, if our ritual is thrown off, it can really throw us for a loop. If you're getting ready for work and you don't have the coffee you normally do, whoa! Look out! We're gonna have a bad day!

Charles Duhigg, in his book, The Power of Habit, describes the constant and persistent routine that World-Record-Holding-Swimmer, Michael Phelps had for himself. Michael Phelps didn't have to get nervous for Olympic races. He wouldn't treat Olympic Races any different from routine workout races, because he kept his routine the same for all of them. Doing so kept his nerves and anxiety in check. It was just another day of swimming.

Once, I was searching through the cheese section at a San Francisco grocery store. There was an older gentleman fuming and steaming and swearing to himself, just loud enough for me to hear him, about how they didn't have HIS cheese! "How could they be out! They're NEVER out! I can't believe this! What on earth am I going to do now?! This is RIDICULOUS! This has NEVER happened before...." and on and on. He kept staring at the empty spot in complete and utter disbelief.

I tried leaning over as politely as possible and suggesting another brand of the same type of sliced cheese, another major company similar to what he was looking for, and even gave my recommendation that their cheese has always been good when I've tried it.

It was like I had just derailed a train. The look on his face was incredible in its confused surprise. Like I had just suggested the most revolutionary thought he'd ever considered. He picked up the suggestion nervously and began reading the packaging. I walked away wondering how long he'd been using that routine cheese to have it so ingrained in his mind.

I have a routine I go through when doing my dental exams on patients. I try to change it every couple of months, just to keep me from diving too deep into autopilot, but, really, it hasn't changed much over the years. I'll probably be doing it in my sleep long after I'm retired.

There are several principles of influence at play in the use of rituals that people may use to manipulate you, or help you to do good things. If I ask for a smaller commitment, you're more likely to agree to a larger commitment later; a ritual is begun. If you can tag a new ritual onto an already stable one, you can add more habits to your existing ones, like tagging on some unrelated item to a law that's getting pushed through Congress. It's like getting two laws passed in one!

If you want to add exercise to your morning ritual, leave your exercise equipment or gym bag in the way of your usual path, so you practically trip over it. You're more likely to stop your current ritual, and add the new one on. This principle is used to build a healthy and stable lifestyle. It's used to develop new habits, and it's used by your brain to get an oxytocin-dopamine shot that feels oh-so-good, even if it's not always good for you.

This is how bad habits form, but we can also use this to stop bad habits and addictions. Put something in the way that will derail your ritual in a good way, get your brain to snap out of it for just long enough to make a clear and conscious decision about what you're doing, and follow that derailment.

If I were to tell my younger daughters, "All right! Time for bed," while they were zoned-out watching a movie, it'd be rebellion and riot. That goes for my older kids at times, too. But if I tell them, okay, go brush your teeth and put on your pj's, while I pause the movie for a break, they might grumble a bit, but they'll usually jump right to it, then race back to watch another 20 minutes of Netflix or Youtube or whatever it is. Then they know that soon it'll be time for prayers and bed and it's less of a surprise. They know the routine, whether there's entertainment involved or not. There's a ritual and small commitment involved. Then, actually getting into bed isn't such a big deal or leap for them.

Because of rituals, often, your mind will act out in addictive behavior, whether you want to or not. Once the ritual has started, it's incredibly difficult to stop, and often those rituals start with normal, innocuous, seemingly innocent behaviors that wouldn't get anyone's attention, even your own attention. Then, like a balloon being filled with air, a breath at a time, that balloon fills and fills and fills, POP! and you are openly indulging in behavior you didn't want to behave. The dam has burst open, the thoughts —like water— are flowing, and you are not going to be able to stop it on your own until the stream has run dry or some external force wrenches you out of it.

What becomes a large part of many addicts' rituals are dishonesty and lying, being sneaky, or hiding things. Once dishonesty and hiding have become a part of the ritual or problem, or bad habit, addiction is close by. Being sneaky to indulge in selfishness is what isolates us and drives us further into the behavior.

Honesty is the grasping and using of reality. It is standing on solid ground. Lying, sneaking, rationalizing bad behavior, and cheating is like using a balloon to hammer a nail. It might look like it's working, and maybe with enough swings and taps the nail might actually stick to the wood, and everyone around us, including ourselves may think we are doing the job correctly, but there's no way that nail will hold. Once weight is applied, it will fail, often catastrophically.

Using reality —being honest— is what makes the world function. Societies rise and fall on their ability to courageously be honest with each other. Dishonesty is extremely contagious and cancerous. When we belong to an industry where lying and cheating has to be used, not just to succeed but even just to make ends meet, because of their prevalence among all who work in that field, we are in serious trouble. Much of the recent Recession that affected the entire world was caused by dishonesty, by even just a minority of our society.

Honesty takes courage and bravery. It requires facing the world as it really is. It requires a solid understanding of reality. Dishonesty is a fantasy land. It crosses the wiring in our brains. Cross too many of them and we're more likely to have major emotional problems later when we can't reconcile all the fake realities we've created for ourselves and those around us.

I'm not talking about a sincere trust in bad information, I'm talking about distorting reality so we can one-up our peers. So we can look good, make money, so we can hide our faults and mistakes. No one actually improves long term by being sneaky or hiding what really exists. Truth rots when it is hidden and it becomes REALLY, REALLY smelly and infectious until you let it out in the open again. Even if those around you can't tell what's going on behind the scenes, they can still sense that something about you is "off" and not quite right, and eventually it will be discovered. The longer it takes, the worse it gets for you and the rest of the world.

Many know on some level that what they are doing isn't enjoyed by our loved ones and so we hide it, either out of rebellion, shame, or to get more of it. People will lie to doctors or parents about drug use and abuse. Spouses will sneak and hide their pornography. Loved ones will stay away to hide their drinking or cheating. One of the worst things for an addiction is for it to get placed out in the open and talked about. This isn't true for all, of course. Some addicts are happy to act out right in the open, justifying that it's an acceptable behavior, or will be by their openness, and they are just being "true to themselves," whatever that means.

It's easy to act out an addiction around other addicts who are participating the same as you, but it's harder to do it with those who are hurt by your behavior. This is just one reason why support groups are so helpful. Terry Crews, the football player and actor who also made many of those funny Old Spice commercials, is fairly vocal about that fact. "The more you talk about it, the less power it has." The more light you shine on it, the more it shrivels up. (DOUBLE-SUN POWERRRR!!!)

As a caveat, talking openly about your problems doesn't give you justification to go indulge in them, while telling yourself, "It's okay, I'm working on it, and I'm not hiding it, so it's all good." Or, "My family knows about it, I'm not lying to anyone, I'm not breaking trust, I can do this. I'm in control." This is a lie to yourself and becomes a ticket to indulge in it all the more, even though your loved ones still don't want you to, and even when doing so hurts them.

Once you start lying to others and yourself, once you start sneaking it, it becomes almost automatic to rationalize and justify why you are doing it in the first place. Sometimes we don't "hide" it, but we retreat away and distance ourselves from our loved ones and friends, or those we are afraid will disapprove of our behavior. Whether they actually disapprove of it or not is completely irrelevant. If we are afraid that people will disapprove of us, we assume they will, and we become self-conscious, embarrassed, or ashamed, and we stop involving those people in our lives.

We cut ourselves off from those who would help us or those whose presence would help us be healthier. We wrongly tell ourselves that those people are actually destructive to our well-being and progress. Some of those people *may* actually be destructive and harmful to us, so distancing ourselves from loved ones can be really good, or really bad. Often, the reason we are feeling the need to act out in some addiction is because of the poor relationships we are finding ourselves in.

Your brain forms this disconnect between two different desires, two different wants. You'll try to find some way to have both, your family AND your addiction, your health AND the drug. They are opposites. You cannot have both forever. Like trying to serve two masters. One will eventually win out, and unless you make an intentional effort to save your family or health, the addiction will be the winner. Until one wins out, you will be living with a half-starved family, physically or emotionally, and a half-starved self, physically or emotionally, or both.

The second half of this book will give ideas, suggestions, and strategies for overcoming all this stuff, all these obstacles that keep us from getting better. For now, let's talk about what kind of equipment we already have that lets us overcome it.

6

Your Outer Cortex: The Shell of Your Brain That Makes You Look Good

Often we get stuck between wanting healthy relationships and the rewards of them, against wanting to indulge in some destructive habit. Breaking this pattern between two pleasures, two wants that are nearly equally desired by your inner, reptilian brain, requires an outside force. Some external resource. There's a few of them to choose from. The first one is still internal and is built into every healthy human being from before birth, but outside of our instinctive brain. It's the outer layer of our brain, our cortex. Especially the front and side areas, known officially as the Pre-Frontal and Temporal cortices.

This is a very well developed part of our brain that's somewhat unique to humans, at least in how efficient it is, not so much by size. All of the outer parts of our brain are called the Neo-Cortex. Neo for New, because it's relatively new in evolution compared to the incredibly old, probably-similar-to-dinosaurs'-mid-brains and the more hormonally and reflexively controlled lower parts of the brain.

This neo-cortex is like having another person inside our head. It's responsible for most of our "higher" reasoning. It is our "conscience" like Jiminy Cricket to Pinocchio. I call it "Higher," because it's better than that primitive lower thinking done in other, less-fashionable parts of our brain. Actually, it's less hormonally and instinctively driven, and more memory driven. It has the ability to take a great many more things into account including your conscious and unconscious memories that are deeply or not-so-deeply rooted in your brain.

It's like having a secondary, detached, personality in your head. When folks have had an injury to their head or brain damage of some kind, and it affects the outer part of your brain, especially towards the front, it's common to see some major personality changes. I had a friend as a kid who got into a dirt-biking accident. Before the accident, he had been a quiet, reserved, soft-spoken kid. Now he was now gregarious, outgoing, and even a bit reckless in his behavior. Almost like he'd had a couple of drinks, but all the time! With the damage to his outer cortex, his inner and mid-brains were less inhibited, socially, and he became more hormonally driven.

Another friend was in a car accident, and among damage to the front of his brain, he also lost his sense of smell for a time. Some of it recovered, but he could also point out and recognize that he was much less patient with the people around him. He was short-tempered and was constantly working on keeping his patience. Before his accident, he'd never been short-tempered.

It's a curious thing that much of our personality is controlled and driven by the outer-front-ish part of our brain. It brings up the question, again, who are we, really?

When I meet people who are happily acting out their addiction, openly, they like to claim that this is who they REALLY are. This is their true self. They're no longer suffering under the repression of societal norms. They are going to act out their impulses and everyone else can just go kiss their.....

Usually there's strong language involved in explaining their new found freedom.

Who are we? Do we need to go on European or south-east-Asian backpacking trips, or disappear into the wilds of Alaska just to find ourselves (cultural experiences aside)? How about exploring and "freeing our minds" and "boosting our creativity" with hallucinogenic drugs or even milder drugs and substances? What about ditching our family for some more-attractive partner who "truly fills our desires" and "makes us feel complete?" What about no longer fighting against and suppressing the rage and anger that's haunted us our entire life? "I'm tired of holding it all in!" we hear, usually as a family and other relationships fall apart, with a wake of damage left behind.

So who are we? We are a few different people, inside of our skull. We all have multiple personalities. We all have multiple ideas, and philosophies competing for our attention and trying to drive our behavior. This is because our brain is so large and complex and each section is trying to make sense of the information it has and tell the rest of your brain what to do, most of it completely unconscious to us. Even within the "conscious" portions of our brain, we have warring factions and ideologies. We have conservative ideas and liberal ideas competing for our attention.

"Should I eat that brownie on the table? Yes because it's soooooo good, and you're hungry, and it's what, 200 calories? My diet and my gut can afford the extra, this time, and plus, I'm at a party with lots of other people, and it'd be great to talk with at least a few other people about how good it was!"

Another part of your brain says, "No" because, "I'm trying to lose weight for the beach vacation coming up. While it's a good brownie and full of sugary sweetness and chocolate, I know it'll give me some heartburn, and it doesn't taste THAT good to be worth those extra calories." What's a person to do in that situation? Like Democrats and Republicans fighting with each other inside your mind about whether the indulgence is good or not. And most of this happens unconsciously and automatically. Whichever side we've given more attention to in the recent past is the side that will probably win out, unless we've given TOO much attention to that side and we're feeling guilty about it and know we need to back off.

There's also the "conscious" and outward aspects of our brain. I've seen some people list as many as six different personalities residing in our brain, driving our personality. I tend to favor the idea of three: 1. Who we THINK we are right now. 2. Who we WANT to be and are working towards, and 3. Who OTHER PEOPLE think we are based on our outward actions and words.

The first one is fluid and ever-changing and our perception of who we are depends a great deal on our current state of mind, and at times doesn't reflect reality. The third one is partly in our control and mostly out of our control, so it often doesn't do us too much good worrying about that one. The second one, "Who we are deciding to be or become" is, in my opinion, the most important. It's the only one that we can actively decide and take action on. It involves a great deal of trial and error, learning and education, and willpower and determination.

7

Self-Determination: You Ultimately are Who You Decide to Become

The things we want to do, the destructive behaviors that hurt ourselves and those around us, are all superficial attempts to fill some void in our life, or to satisfy some addiction or compulsive behavior, or animalistic instinct. They are all based on survival to some degree. I would argue, from my own opinion, that WHO we are is a mix of who we want to be and what behaviors others see in us. What we DO dictates our personality to others. It doesn't matter if we think that we're nice and loving. If we are outwardly abusive, then others will see us as abusive.

To others —to those people outside of our mind— our actions dictate who we are, entirely. Some of us have met those who suffer from delusions of grandeur, those poor folk who think they have more money than Bill Gates, but they're living on the street "just out of humility." We think they're just a little "off" and rightly so. Who are they? A homeless person who cannot grasp reality? Maybe. Maybe not.

When I was in college, one of the run-down apartment complexes I lived in had a pool and a hot tub in the center quad. Fancy, right? One night, a date and I went down to relax in the well-lit hot tub in the central courtyard of the complex, and there was this older-middle-aged guy sitting in there. Picture, a middle-aged, balding guy, who's letting what little hair he still has grow long past his shoulders, wearing a great big fake-gold chain necklace, with a big ol' beer belly, hanging out in the hot-tub of an apartment complex filled with college students. Weird, and a bit creepy. I guess he was trying to impress the college students like he was still cool, you know, because he snuck into a complex full of college students. He wouldn't stop talking about his "old buddies" that he had made rich and famous. "You may have heard of some of them, Harrison Ford? Oh, you HAVE heard of him? Yeah, I helped him get his first job. I gave his buddy George Lucas the idea for Star Wars" Sure. Sure. We just smiled and nodded and didn't hang out there for too long.

When I was living in Vancouver, Canada I spent an hour talking to a homeless guy on Robson Street who was trying to tell me about his inventions that would revolutionize the world, "but the whole system is rigged, and other people just steal my ideas. No one has any original ideas anyway, they're all standing on the shoulders of giants and taking credit for it. But my ideas? All my own." Perhaps. Perhaps. How I saw this guy, is definitely not how he saw himself. He tried to convince me that his idea of a multi-winged plane with kite-like fabric wings would change the world. I tried to carefully remind him that his knowledge of wings, kites, and Bernoulli's principle were ideas that others came up with. Nope, he says, he came up with them all by himself.

Another great majority of our personality is Who We Want To Be. Which direction are we headed? What are our goals? This is who we are on the inside. Sometimes others will be able to see this side of us, and that will contribute to our outward personality, but mostly this side of ourselves is known only to us. A migrant farm-worker who spends 60-80 hours a week working to provide for their family and eke out just a simple survival-based living could be either someone who gets hammered-drunk on the weekends or someone who takes night classes and reads self-improvement books. Neither person may actually escape the poverty they are in, but for the first one it won't be surprising if they can't escape poverty, and for the other, we'd be surprised if they or their children couldn't. The hammered-drunk may wish he could escape the alcoholism, but may not have a social support system, or any knowledge of how to escape, and no tools for doing so, no matter how much they want to inside their head.

So, who are we?

It's mixed. We are an amalgamation of how we act, what we are working towards, and how others are seeing us. Like the primary colors of our eyes that put the whole world together in amazing scenes and views, we are complex, complicated personalities, much of it driven by our environment. There are other ideas out there, but I think these are what matter most. Fortunately, as humans, we have *some* ability to change our environment.

Some argue that it is their passions or their interests that define who they are. This is how others see us. But hobbies and passions, quirks, and lifestyles, these can change as we grow up, mature, age, move from one city to another, move in and out of various circumstances in life. Physical ability and attractiveness may form part of how we define ourselves, especially when we are young, and these may become lost as we age. Friendships and careers will shift and change as we grow, without any loss of individuality. I used to define myself as a Californian through and through. I made every decision in my life so that I could stay near the coast of Ventura County, California, near my very large extended family and near the ocean and it's beauty. A great many things combined to push me to the Mountain West area of the United States, 800 miles from the nearest ocean. And, I'm loving it here. I came kicking and screaming, and now I'm trying to convince others to come join me.

Our interests are not who we are, though it is how others will see us. I think that what matters most to our person, is who we WANT to be. What was more important to me than being near siblings, parents, aunts, uncles, cousins, nieces, nephews, second cousins, the pleasant weather, tons of fresh produce, the ocean, sailing, surfing, diving, fishing, exploring, and everything, was to be a good Dad to my older kids. And to do that, I had to leave everything else behind. And move to an area where my field of work is one of the most difficult and competitive in the nation. To me, it's still worth it. The difficulty of being here has pushed me to grow in ways I never thought existed. It's been a discouraging, frustrating, fun, and exciting adventure.

Once, while volunteering as a missionary, I laid awake one night, imagining music in my head. For a few years, I had been wanting desperately to write music, to create majestic, amazing symphonies, to rock out out on a stage having fun stirring up a big crowd. That night, a song popped into my head that was more beautiful and amazing than anything I'd heard before. I wanted to get it down on paper. I knew it would take a big time investment to do so and I couldn't do it while on this volunteer trip. I had to make a decision, quit the work I was doing to try to capture that ethereal melody, or ignore a dream to stick with the earlier decision I'd made. It was a tough one. For that time, I decided to stick with the trip.

The next day, the music was gone. I couldn't get it back or even remember how it went. I was disappointed and sad at losing out on a creative opportunity like that. I needed to find a way to get the music out of me or I thought I would burst, but I couldn't even remember how it went. I needed a creative outlet or I would be miserable forever! And then a thought occurred to me: Raising a family is probably the most creatively demanding endeavor ever posed to humanity. It would take all of a person's imagination, cleverness, persistence, and fortitude to accomplish! I would have to grow to match it.

Are we getting better, stagnating, or getting worse? I would argue that stagnation and getting worse, are basically the same thing. Plants often cannot stop growing without stunting or withering in some way. We are the same. Upward, or backward are the two choices. Even if a maximum size has been achieved, a plant must keep putting out new growth or it will die. We have to do the same as well. I have met too many people who retire, just to spend most of their days in the armchair watching TV. They might get out to garden or do something once in a while, but it doesn't take too many years before that armchair becomes their prison and they can no longer get out, even with many, many years of life still ahead of them.

Perhaps they started retirement in the right direction, with traveling, trips, and golf, and fun outings getting them out of bed each morning and out the door. Once they've gotten their fill of that, it eventually becomes easier to sleep in a little longer or postpone that outing a few days. The TV becomes easier than gardening, and slowly the armchair becomes their prison and their only window to the outside world is that screen. Many studies have shown that those who have obligations to other people, neighbors, family, or church tend to stay healthier and live longer. Being committed to other people, in some form, is necessary for life, and to do that, we have to find ways to keep improving and keep up with them.

We live in a river-current of life that happens to everyone. Success and tragedy happen to everybody, some in great swooping crashes or exaltations, some in slow creeping growth. Sometimes both success and tragedy occur simultaneously. Very often, it's outside of our control. Like being caught in a fast-moving river. Imagine a bucket of water with a few holes in the bottom. All the water wants to get out through those tiny holes and achieve the success and freedom of the open air. Only the water closest to those holes, or through luck and currents of the bucket, will find their way out of those holes, no matter how hard all the others try.

Perhaps some careful finagling and scheming and hard work will have some influence to us achieving Outlier status and incredible fame and glory. Most of it, though, will be due to the circumstances we were born and raised in. Of course, none of us will ever know what we have control over until we give our life to gain it. Perhaps we could have changed things, maybe we couldn't, and we'll never know unless we tried and tried and tried.

How we deal with these successes and failures is what makes us who we are, and is also contingent—or based—on who we've been and what life has handed to us. But with a frontal lobe, like we've got in our brain, we have a very unique natural gift that lets us choose which direction we are going to go or at least work toward. Are we just going to let ourselves get swept downstream, over waterfalls and smashed into the rocks? Or are we going to fight and kick and swim to either get upstream or at least get out of the river for a break? Sometimes, going with the flow is the right thing, sometimes, it's not.

So, what is it that we should want in life? Abraham Maslow argued that we first need Food and Water, then safety or shelter, then love and belonging. When those needs are met, we can learn to feel good about ourselves, and even improve ourselves. This is known as the Hierarchy of Needs.

It's drawn like a pyramid. At the bottom are Food and Water. Above that is Safety or Shelter. When that has been met we can focus on Love, Intimacy, and Belonging. With that fulfilled we will develop Self-Esteem and finally, what Maslow called, Self Actualization. Self Esteem is the respect we give ourselves and the respect others give to us. It is the confidence that what we are doing is right. Self Actualization is the ability to be at peace with yourself and express it to others through our morality, creativity, and problem-solving. It is acceptance of the world as it is, and the acceptance of ourselves as we are, and then the improving of both and finding meaning in this.

Maslow would argue that no one is going to worry about improving their talents if they didn't have enough food or shelter. But this isn't always true, oddly.

We've heard stories of prisoners in awful POW or war camps who find ways to express themselves artistically while they starve to death. We've known of people who've resorted to cannibalism to survive a cold winter or famine, while others in the same situation didn't. We know people who resort to crime while their neighbors or even siblings don't. Why do some people stay civilized in the face of death and suffering, while others revert to crude animals?

While Maslow's pyramid holds true *most* of the time, it's not guaranteed, and things are often found out of order. How many people do we know that have all the food, love, shelter, and talent that anyone could want but don't have any self-esteem? Or those who have anything anyone could ever want but still feel unfulfilled or even lonely? What about those who don't have food, or shelter, yet can be friendly, helpful and seem totally fulfilled? There's some Zen Buddhism going on or something.

Maslow didn't have it wrong. It's just that when it comes to needs being met, we don't actually need those needs met before we can move on. We simply need hope for those needs to be met. We need faith and confidence that we'll get what we need. A Hope for the future. A certainty of some kind, or any kind, that your effort will be worth the sacrifice. Worth the effort. We will happily give up food and shelter if we know we will get it later. We will happily give up love and affection if we expect it to come sometime. We will happily wait for anything if we expect to obtain it someday soon.

If we expect self-esteem and connection to be met someday, we can put it off while we work on self-expression. If we know that food and shelter are simply around the next corner, or maybe even at the end of a 4-month through-hike of the Appalachian Trail, we will work on self-actualization in the meantime. If we are fully confident in finding a partner someday, we can work on improving ourselves.

Tony Robbins, the world-renown speaker, preaches about another set of needs that every human has. He claims that every human needs:

1. Certainty or Security.

2. Variety to spice up our life, because security gets boring.

3. Significance. This can be translated to uniqueness or individuality.

4. Love and Connection, we need relationships, (note that he lists this after the significance/individuality).

5. Growth. This is a special one that we can live without, but helps us feel fulfilled. We need to keep growing or we are headed towards an early grave. There is a 6th one, but I'll save it for later.

Yes, Maslow's Hierarchy of needs are important and the order of needs isn't wrong. We won't hope for self-actualization if we don't even hope for survival until our next meal. But the need itself isn't important, what's important is the hope we have or the expectation we have of that need being met.

What's clear is that we can't rise any higher on the Hierarchy of Needs than who we've decided to be, no matter which order we go in. Often, we can't do this alone, we have to surround ourselves or fill our head with people we admire. Who we admire is who we will become.

When we look at organizations such as the Delancey Street Foundation, we see this in action. Delancey Street began in San Francisco and has grown to several locations across the US. Residents of the center come from jails, rehab facilities, homeless shelters, or walk in on their own. Most have a serious criminal record or crippling addiction. They are missing MANY of the needs Maslow says we all require.

By the time the residents leave 2-4 years later they have an equivalent of a high school education, between 1-3 marketable skills, a job, and a new life. All without any paid experts, grants, or government funding of any kind.

The Delancey Street centers give their residents some basic survival-based rules: No drugs, no alcohol, no violence, and no threats of violence. Then each resident gets assigned to another resident who watches over them, mentors them, and makes sure they're keeping the rules. Each person is a teacher of someone else. They start off by learning basic job skills, like showing up on time, using personal hygiene, etc, and grow from there, often being mentored by someone who's only been there a few days more than they have, and usually these pairings of mentor and mentee are rivals from a gang or race that had opposed or fought each other before life in the program. Most residents have been addicted to hard drugs, been involved in prostitution or gangs, been in and out of jail, and often been homeless for long periods of time. While they sometimes had the basic necessities of "life" before they joined Delancey Street they had no education teaching them how to take care of their own life. They couldn't progress any higher than survival-type living. No healthy relationships, no healthy habits, no responsibilities. No idea or knowledge of how to live beyond a meager subsistence.

In Delancey Street, they are taught these basic life skills. Yes, they are granted the gift of shelter and food and clothing, but most importantly they have a society of other people that they can fit in with and grow with. They have finally achieved some safety, some love, and some belonging, and once they get and understand and feel trust in those things, they can think clearly enough to grow. To achieve self-actualization.

Compare that to all the countless through-hikers on the Appalachian Trail or Pacific Crest Trail, or Europe/Asia-backpackers. Those who were born and raised with a great gift of knowledge, home, stability. They decide to leave it behind for long, extended periods of time, to face exposure, hunger, danger, and the like, in order to pursue self-realization or self-actualization. Sure, most of them know they can just go home and get those basic needs met again, but why leave them in the first place?

What I'm trying to point out is that having food and shelter and basic needs, doesn't automatically lead to earning, receiving, or obtaining the other needs, though they are very important.

There's definitely more to our story than just an order of needs, and this order changes from culture to culture and even throughout the life of a typical human being. The Doukhobors who immigrated from Russia and Asia to British Columbia, Canada developed a reputation for burning all their worldly possessions —all their clothes included— to prove their detachment from worldly belongings. Other cultures will push their children to join one of the medical or legal professions, assured that material success is the path to self-realization. Some people raised in poverty will sacrifice all personal relationships to achieve financial security, while others will ditch their large homes and stable careers in the city to build an off-grid tiny-house in the woods.

"Life" happens to all of us. And by "life," what I mean are the events that cause us emotional scars, or priceless joy. Children, marriage, breakups, divorce, health, injury, illness, success, poverty. These all have a lasting impact on our life and what we will want or get out of it. Things happen to us that affect how we think and even our ability to think, either by hindering it or improving it. And this level of ability to reason and think is what dictates our needs and wants and how we interact with the people and world around us.

When we have some ability to determine our lives and our actions, and we gather around others who have a similar determination, there's no end to what we can accomplish. Humanity, as a species has a collective drive to survive and thrive. When we can align with each other towards a common goal, nothing will stand in our way. This can go both ways, good and bad, or both at the same time. Rarely is something we decide to do solely "Good" or entirely "Bad." Everything comes with a cost. When we are clear headed and respectful to ourselves and others, we tend to make decisions that are more good than bad. When we are not clear-headed, we tend to make decisions that are more destructive than helpful. This applies collectively and individually.

So what if we could get most of us clear-headed and then find things we agreed on and all worked towards those things? The trick is, presenting information to each other in ways we can all understand, and being in the state of mind to receive that information.

If we could do that, I doubt there would be many needs that any of us wouldn't find being fulfilled.

8

The What: The Food and Intimacy That Keeps Us Alive

Addictions and destructive behaviors arise as a way to fill our needs. Usually as a substitute to get us through some trying or stressful time. What can happen, though, is that the addiction hijacks our biology and prevents us from enjoying the "real thing" when we can get it again. Louis Zamperini described seeing a fellow POW in his camp who became so stuck on cigarettes to curb his hunger that he refused food when it was finally given to him.

Think of a husband or boyfriend (or a wife or girlfriend) who's so entrenched in porn that he doesn't give any attention to his wife or girlfriend, even when she wants to be intimate. Or an old woman who's so bitter and negative she can't enjoy a beautiful national park or an amazing sunrise without complaining about how the clouds could be better, or if only there weren't so many other people ruining the view. What about the child who's so stuck on a video game, they would rather stay inside staring at a screen, playing with strangers, than playing outside with friends? What about someone who's entrenched with feeling depressed and lonely but can't crawl out of bed when friends actually want them to join them somewhere?

There's an incredible amount of reasons why anybody can resort to addictive behavior. The reason and the behavior may not even be related, but fortunately, they usually are. Somehow, somewhere an imprint is made on our reflexes, our instincts, and it sticks. Like ducks following the first thing they see after they hatch, somewhere in our development we get imprinted to some behavior that appeals to us.

Since we are always developing throughout life, we are always at risk of a bad imprint. Some are imprints that we foolishly or naively chose for ourselves, some are put on us without our knowledge and consent. It doesn't matter too much whichever it was. The result is the same: we're drawn towards something and we may not know why.

However it comes about, the responsibility lies on the affected person to overcome it. We usually need some external help, outside ourselves and our big brain to do so. We will need more than ourselves to overcome some problems, regardless of whether or not it was our fault to begin with. No one can force us to change for the better, we will have to decide or want to change, ourselves. The actor/singer Will Smith does a great little video about the difference between something being your or someone else's "fault" and still having the "responsibility" to do something about it. Our addictions may or may not be our fault, but we are still responsible for them.

Sometimes we are fortunate to find a way to live with and be able to act out our imprinting in a safe and healthy way, or at least, a sustainable way. We find a partner who'll happily and supportingly indulge in our imprints (without letting us act out in destructive ways!!), purely for our sake and their love for us. That's a great thing.

If you were raised in a tribal setting, eating little but grains, nuts, fruits, and a few wild animals, you'd likely be satisfied with that the rest of your life. The starkly different modern diets of many countries would be incredibly upsetting to your body and would take a great deal of time to acclimate to. The reverse is also true.

If you were raised in a modern society with hamburgers, ice cream, and all the cheap chocolate you could get for Halloween, going to a "Paleo" type diet is a major life change and often one that you're never truly satisfied with. You need those cheat days every once in a while to indulge in a good donut. It'd be easy to live a paleo diet surrounded by others eating the same foods. It might be extremely difficult to live it when your apartment is above a fast food restaurant, and you smell those delicious french fries often. Mmmmm.

If you were raised in a conservative religious home, your ideas about life will be very different from someone raised in a home without any moral restraints. Sure, you might swap some stories and ideas between each other, but not usually in a favorable light. A former Seventh-Day Adventist may always feel a twinge of guilt and shame each time they partake of meat, and a newly-converted Mormon may always feel that pull towards a nice cold beer. Those imprints are hard to shake. The same is true of sex and intimacy.

We wouldn't judge anyone who goes from a primitive Paleo diet to a Modern Diet, we'd expect it to happen. We praise those who can go from an unhealthy modern diet to a primitive one, and never think less of them for having a cheat day once in a while. We should also never judge someone for moving from conservative values to more liberal values and morals. Many people would celebrate that move and the "freedom" that person has achieved. We tend to expect it when someone is surrounded by it. Conversely, we celebrate it when people leave an amoral lifestyle and place several new Conservative boundaries on their behavior. We usually want to be supportive of them and their efforts. Generally, we love to see people make positive changes to their lives if we know how difficult it is to do.

There is a great painting by James Christensen, titled, "Fishing" that portrays a humorous figure sitting on a stool in a gray room, with a fishing pole and the line hanging down into a small pail. I always treated it as just a funny picture until someone pointed out some deeper meaning to it.

This character, as fictional as he is, must've started his fishing hobby the way that anyone does. On a lake, or river, or ocean. For the purpose of actually catching a fish. The thrill of the fight, the fresh air, the food, the peace, and tranquility might have all been good and legitimate reasons for beginning the pastime. Yet, over time, the monotony of waiting for a fish to bite became the norm, and the waiting became the expected outcome of the ritual.

Eventually, it didn't matter whether there was any fish to catch, or any fresh air or scenery to enjoy, or even if the food from the catch was needed. He fished simply for the act of "fishing", of holding a rod in hand next to any body of water. Now he sits alone, for hours at a time, staring at a pail, accomplishing nothing, except satisfying his mind and body for what they are used to: sitting and waiting. His life changed, slowly, over the years, but not in a positive way. He took the path of Least Resistance and Less Effort, until his life isn't accomplishing much of anything, for himself, or anyone else. Taking the Easy Road is really nice sometimes, but it can't be the road we stay on.

Maslow came up with a set of needs that healthy humans *need.* I just spent too long calling into question the importance, order, and requirement of each need, but the gist of the whole thing is that we are biological creatures and we need certain things from our environment to satisfy biological needs. Our DNA demands and expects it. But, our human DNA has checked and balanced itself (or improved itself?) in giving us a large brain that can see past and beyond these needs, some of the time. Not even very much of the time, but that check to its power is there just enough to keep a highly complicated human creature on this planet, as successfully as possible.

We do need food. We do need shelter. We do need a society to belong to. We do need close friends and loved ones that we can form deep emotional relationships with. We do need stimulation for our mind, and to be out of pain. We NEED to learn. All addictions and destructive behaviors exist because of and originally to meet these needs. Either alone or in some combination.

The need for material possessions, shelter, and security, can become addictive. The need for companionship can become addictive. The need for sex can become addictive. The need for peace and tranquility, for self-esteem, can become addictive. The need for love and attention can become addictive. The need for food can become addictive.

In short, a single biologic impulse or need, can start to trump and overpower all other needs and result in your demise. Have you ever made a simple fly-trap, or bee-trap, or hornet trap? They all work like this: A container with a small narrow opening contains some sort of food or imitation food that attracts the insect. The insect smells this food and flies into the opening. Once inside, whether or not there is food, the insect can not or will not leave, because that small opening leading to their escape does not smell like food, and they smell food, so they are going to keep searching around for that food.

A super-simple fruit-fly trap is like this: Some lemon juice or vinegar placed in a small dish or container. Some clear-wrap is placed covering it, and a small hole is punctured through the clear plastic wrap. The fruit flies will smell the fumes from the dish, track it to the hole, crawl inside, and never leave. Usually, they are killed by the acidic vinegar and their bodies dissolved. Gruesome, but it works so well.

This often happens when we try to open a window to let a house-fly out of our home. The fly is hanging out on the glass, trying to get loose, with its obnoxious buzzing and bouncing off the glass. But once we open the window, the outside air doesn't smell nearly as much like the food it wants, compared to the air where it is already hiding. So it leaves the window and starts buzzing around the room again. So frustrating.

Food can be an addictive substance by itself and I'll use it as a detailed example of how a need becomes an addiction. Buckle in, you're going to get a crash course in Nutrition and Diet! Hopefully, much of it is already familiar.

Our body needs fuel to survive. This fuel comes in the form of Carbohydrates first, Fats second, and lastly, but no less importantly, Protein. Think of it like this, Carbohydrates are like Gasoline. Fat is like diesel, and Protein is like wood. We can make car and truck engines that run on any of those fuels (yes even wood, check it out), but gasoline and diesel work the best. Gasoline-like carbs are quick and efficient. Sluggish, but ever-so-dependable Fat is like diesel and also accompanied by some funny smelling by-products (ketone bodies). Burning protein for energy is like running a car on wood, or wood-gas. It works, but it's clunky, slow, really inefficient, and eventually, the soot plugs everything up (unless you're flushing it out with a gallon of water every day. I'm looking at you Keto Dieters!).

"Sweet" food is made out of carbohydrates. Food made out of carbs is generally sweet on our tongue. Some carbs taste sweeter than others. Carbohydrates are made out of various sugars, usually glucose, but it could be any of a bunch of different sugars or combinations of sugars. Glucose is the most abundant organic molecule on this planet, which is why it's also the primary source of fuel for all animal life on this planet. Glucose gets attached to other things in all sorts of different ways, or linked into big chains, or woven into huge webs by the various organs of our bodies. It can get packaged and stored in all sorts of ways by various plants and animals. The simpler it is packaged, the sweeter the taste and the faster our body can digest it. Sugar is scientifically known as sucrose (among many other names) and is a glucose and fructose stuck together. They fall apart pretty easily and are very sweet and easily digestible.

Lactose is the carbohydrate found in just about every kind of milk and is made of a glucose and a "galactose" sugar stuck together. Maltose or malt sugar is two glucose rings stuck together. Starches are webs of glucose that plants use for storage. In animal muscles and livers, woven sheets of stored glucose are called "glycogen". It takes different enzymes and machinery to breakdown and use each different kind. Some humans can't break apart milk's lactose as they grow up from childhood. Their bodies quit making the enzyme keys that unlock that type of sugar. The germs in their stomachs have no problem with it and they will gobble it all up, giving off lots of gas and abdominal discomfort as they digest it. This is what we call "lactose intolerance."

Wood is a real tough carbohydrate made out of glucose, but we humans don't have the enzyme machinery in our stomachs (or mouths) to break it down. If our stomachs find wood, or "dietary fiber" in our food, it gives the signal to our intestines to keep things moving down the pipe, efficiently. So, even though it's not a source of fuel, very small amounts of wood and fiber is important to our health. It keeps the conveyor belt moving nice and regularly, if you catch my drift. Few animals can digest wood by themselves. That privilege is left to various bacteria and fungi. Cows have four stomachs just to house colonies of bacteria that'll break down the wood for them after they've chewed it to a pulp.

Plants also store glucose for their own energy in a carbohydrate known as "Starch". Starches are super useful. Lots of plants store it in ways that are easy for them to consume, and it's also easy for us to consume. Grains, Roots, and Fruits are glucose storage containers. Wheat, Potatoes, Carrots, and Apples. Cook them so they soak up some water, become mushy, and you've got yourself lots of easy fuel. Mashed potatoes. Mmmm.

Corn/maize is super full of glucose. So much so, that it has become a dominating source of sweeteners, starches, and food on the planet. Corn syrup is free-floating glucose in water, as simple as it gets, though not as sweet as it gets. Fruit beats it out on the sweetness scale.

Fruits are a plant's food storage for fructose, meant to help a growing seed have an abundance of fuel to sprout and grow or to attract animals who will spread the seeds around and help the plant to grow. Like baby Superman's space pod that takes him to Earth. Fruit is the little pod that can be dropped and scattered to help little baby plants inhabit new lands. Fructose is slightly different from glucose in shape and that slight difference makes it a lot easier to soak it up from your food. Glucose has to get through several security checkpoints to get into your bloodstream to be used for food.

Fructose is special, it gets a free pass. No special gateway enzymes and selective channels needed, you eat it and it's yours. Glucose... not so lucky. Fructose was meant to be a seasonal thing so it didn't matter if the security wasn't as high on it. "Ehhh, it'll only be around for another week or two, so just let it in and let it do its thing."

Because of this, fructose became an important way to pack on the pounds and store food in our bodies really fast. Get us through winter and the lean months. Super important. And because it's super important, our tongues think it's super sweet. Much sweeter than glucose. Have you ever seen what most bears eat at the end of summer and in the fall, before they hibernate? They pack on the pounds by eating all the wild berries they can find.

So, what happens when big corporations, little businesses, and hungry societies figure out that with some simple chemistry, we can turn glucose-filled corn syrup into the cheap and overly sweet awesomeness that's known as high-fructose-corn-syrup? We get obesity and diabetes. Oops. Oh, and food addictions. But, it IS cheap and abundant, and since humanity has always been on the verge of starvation for all of its existence until just recently, we don't like to turn away abundant cheap food. Plus, if we have to spend more energy and money to get more food, that's less energy we can spend on other ventures, helping other people in other ways.

What about fats? The Other major source of fuel for our body. Some natural, plant-based fats are like necessary vitamins to your body. I'm talking about the Omega 3's and their relatives. They often naturally have oil-based vitamins mixed in with them like Vitamin A and E. They let your brain develop and work well, they make for good hormone-balance, help with healing, and work for really efficient energy storage and calories. Your body is supposed to like fat, but it also gets overused and abused.

The purpose of popular Keto diets is to starve your body of any carbohydrates so that it's forced to burn up the stored fat you're keeping everywhere. Those Ketone Bodies, I mentioned earlier, are a by-product and alternative fuel used by your body and brain made by chopping the long fatty chains into short little burnable ketone links. In many of us, the abundance of carbs in our modern diets keeps the fat-burning machinery on hold. When we run low on carbs, we get ketosis (breaking down of ketones) going. Our body will start to eat away at that food storage we've been lugging around for too many years. It's a great way to survive a winter or drought when food is in short supply. It does take some getting used to ignoring the warning bells and alarms your body sets off the second it starts burning fat. It goes into emergency mode as soon as it does, warning you that you are at risk of starving if you don't look for new food soon. Nowadays, that's not really true, but those instincts and reflexes are still built into us and were meant to be difficult to ignore, for good reason.

What about all the other stuff that makes food so delicious? All those things that our taste buds enjoy so much? Turns out they're all vitally important to our bodies, too, in certain, natural amounts. Salt? Check. Sour is the tartness that comes from plants loaded with Vitamin C or Citric Acid. Check. Bitter? It keeps us from eating poisonous plants but is also found in small amounts in a number of healthy plants like cabbage and broccoli, the "mustard" plants. We like little bits of it, but we reject large amounts of it.

The last one is Savory or Umami. It comes from our friend, glutamate, an amino acid found in proteins. It's the "taste" of protein. When you taste it, it means you're eating something with protein in it. Which is good. Chicken soup and soy sauce both have lots of free-floating glutamates, and along with it a lot of broken-down, easily-digestible protein. We make concentrates of this flavor by using yeast to break down protein. It's also known as autolyzed, or hydrolyzed yeast extract. Add a little salt (sodium) and you get the same effect that food companies mimic by using MSG, monosodium glutamate. Salty Protein. "How can you go wrong?" your tongue thinks.

Just like we talked about earlier, glutamate is used by the body, especially as a stimulant. So...not only does MSG taste good, but it gives you the tiniest buzz as well.

So let's imagine the perfect tasting food. Sweet, salty, protein-y, and fatty, and maybe a subtle bit of tartness. Bacon comes pretty close. It might be missing the tart so keep some orange juice with your bacon and eggs breakfast. A loaded baked potato. Ranch Doritos. Salted caramel ice cream. How about a really good cheeseburger? Mmmmm. I think that's got all the basic flavors in there.

When you eat food that's important to survival, not only does your tongue tell you you're doing a great job, but your stomach and your brain are going to tell you the same thing. Eat a super-concentrated version of it and the amount of dopamine skyrockets as well. Have you ever felt slightly intoxicated from eating too much?

Eating is meant to be a social thing as well. Something you do with your tribe, your family. Getting food was always a joint effort between groups of people. You get feelings of oxytocin, of love, while you enjoy the sensation of taste and dopamine rewarding you for a job well done. The feeling that, yes, this is right.

In the end, there are lots of good things that come from eating, in addition to just "living." Our body is designed to enjoy food that is good for us! Remember at the beginning of the book when I said that the second purpose of life was to eat and collect enough nutrients so that DNA can replicate itself and you can have offspring? Food and being social is super important. But when we focus on just the isolated, simple ingredients that don't do much, except taste good, we fill our body with stuff that actually isn't! Like having a quaint town or city, but with no buildings, just facades. All fakery and emptiness. A dietary Potemkin Village.

This is part of the problem with diet sodas and artificial sweeteners. Our tongue and our brain get the message we are eating something that aids our survival, but no such substance actually makes it into our bloodstream. Sweetness with no fuel. Trickery! Lies! Deceipt! So what does our brain tell us to do? Well, eat more of it to make up the difference! So much for the diet part.

On the opposite side, what happens if you can manage to choke down a hearty, heavy, single piece of whole-grain bread, maybe with some all-natural peanut-butter and no-added-sugar jam? I don't know about the rest of you, but after one of those, and some water? I'm good. Pretty satisfied. It was so much work to eat that thing, and it was so energy dense, that, I'm kinda done eating after one of those.

I don't mean to pick on food addictions, or the foods that we all enjoy, but this is a pretty common issue in Western society that just about anyone can relate to. It's a safe example to give for general audiences (until big food companies shut me down for trashing on their products!).

We can find a substitute in this world for anything we're missing. Whether it's human intimacy, stimulation from learning, food, shelter, family relationships, self-improvement, love, it doesn't matter. There's some counterfeit out there or something that'll help us forget we're missing one of those things.

Let's talk about a BIG one, for many, many, many people in this world: Sex. This thing causes a LOT of problems. It's responsible for life for all advance life forms and it gets carried away. Pay attention the next you're at a party or with a group of friends, as soon as something related even slightly to sex and intimacy gets brought it, it takes over the conversation. Everyone's minds will very quickly perk up and have something to add to the conversation about it, or most will enjoy participating and listening to the conversation.

Our brain is designed to focus on it. Thinking and talking about it is very contagious. We try to put limits on public expression of sex because it is so incredibly powerful on everyone's minds, whether we think we are enlightened and mature, or not. If you don't think seeing it, talking about it, or hearing about it has an effect on you, I encourage you to look very carefully at yourself and your life.

Consider someone who's cheated on a partner or spouse, even if it's as tame as checking out some porn, or ogling someone walking by. You can get sex in some form just about anywhere if you're willing to settle for it or risk it. And it is "settling for less" and risky to get it anywhere but from a devoted spouse or partner —though many would argue that getting *something* is better than the *nothing* they're aren't getting from their partner. Well, touche. Still, obtaining an intimate relationship with another human being is supposed to take trust, time, effort, sacrifice, be safe, and require a large amount of vulnerability.

Any false form of intimacy —whether it's from pornography, any type of masturbation, erotic literature, prostitution, the internet, illicit sex, an affair, or just a fling with no commitment involved— is much easier and requires less vulnerability. Yet it's riskier and more likely to be harmful to your health and your mental and emotional well-being. It doesn't take the same time investment, and there's little trust involved. It requires less initial effort and sacrifice. And the rewards of such a relationship are equally diminished.

Often the downsides for false intimacy come later and require much more work to deal with the aftermath. Kinda like if you ate junk food all the time and got diabetes after a decade or two. Over years and decades, your ability to feel and connect emotionally with others will have become handicapped and crippled.

I understand that sex and intimacy are as complicated as I said the human brain was earlier. There are a bajillion reasons why people have some issue with sex or intimacy. It could be because of abuse, neglect, a stressful job, health and wellness issues, the result of bad information or education regarding it, curiosity that became addictive, whatever. The end goal —and it can take years for two people to achieve this— is that the two people in the relationship are working toward becoming of "one mind and one heart" while keeping their individuality and respecting the other's.

They are willing to sacrifice some measure of themselves for the relationship or the other person so that they can build a loving relationship out of two separate individuals. This is near impossible to achieve, but many couples out there are close and anyone can work on getting closer than they are now. True Love doesn't fall in your lap like the movies show. It is built over time. Physical and emotional attraction DOES fall in our laps, often unexpectedly, and can be difficult to deal with when it comes on strong.

Relationships are made a bit more difficult when one person isn't working as hard or doesn't see the need for improvement. Fortunately, a marriage doesn't have to be sent to the trash heap because our sex-life is stinky. Don't throw the baby out with the bath water. If we become convinced that our life is miserable because our sex-life is miserable, non-existent, or even just hard, that *might* —really, might— mean we have an unhealthy attachment to sex ourselves. And we'll have to figure out why.

Perhaps we are jealous or envious of the apparently exciting sex-lives of other people in books, on the internet, movies, or the news. What's the best solution for that one? Stop watching TV, movies, and news, and whatever else is exposing you to other people's private sex lives, out of a full context. Don't throw fuel on that fire.

In our home, we have pilot lights in our water heaters, our furnaces, or our stoves and ovens that are always burning, safely and serenely. A small blue flame whose only purpose is to ignite a bigger fire when we need it. It's only once we throw some fuel on it that it'll flare up into a nice flame for cooking food, warming our home, or if we're not careful, burning our house down. Our bodies have a pilot light that's always on, or at least intermittently, and for the most part if we're not being exposed to sex, if we're not thinking about it, or seeing or hearing others talk about it, our own sex drive calms down a great deal to our Pilot Light Mode, at least once we get through puberty....

Now, when we're in a healthy, committed relationship and we are sharing a home, or room with another person, and we see them in an intimate setting, that pilot light is getting some juice and will wake up real fast in a burning flame. It's almost always near impossible to try abstaining from some form of sex when you live with someone you find even mildly attractive. If you don't end up being romantic with them, your body will take control and you'll end up being "romantic" with yourself, someone else, or something else.

To continue with that, what happens if that flame lights up, but nothing ends up being fulfilled? No deed is done? Ugh, frustration. We gotta go blow some steam, satisfy the urge, or vent our feelings somehow. Maybe we start masturbating, watching porn, dressing in more provocative ways, or losing our temper. Maybe. It's better to find ways to prevent the fire from lighting in the first place by avoiding situations, scenes, thoughts, and imaginations that stoke it. This isn't being prudish or not being in control of our bodies. This is learning to control our body by avoiding the circumstances that will cause us to lose control of our body.

Like others have pointed out, the old "Victorian" ideas of modesty and morality weren't there to be prudish or self-righteous, but to slow down the spread of syphilis and gonorrhea at the time. Any rules we develop as a society for our time will be to halt other destructive problems we are facing, and they will be similar to the measures that countless other societies have taken over the course of history. Of course, turning down the flame, or getting the machinery to slow down to a low idle is very, very uncomfortable, especially when you have to do it quickly.

Sometimes we need to satisfy the urge, just to release the buildup of hormones. We will find some way to climax or orgasm just purely for a hormonal release. That can work for a while. If that's all we're doing, we will eventually feel empty and unfulfilled and have to move on to bigger and crazier things and more intense stimulation. At least occasionally, sex needs to be emotionally and mentally fulfilling, with a loving partner, on equal footing with each other. We need to form a complete mind-body-spirit connection with someone. It needs to fulfill all the needs for oxytocin, dopamine, serotonin, GABA, glutamate, and everything else that lights up during a satisfying sexual experience with a trusting and loving partner.

This is hard to accomplish, but it is what biology is designed to accomplish. There are generalized inherent differences between men and women and this mismatch causes a lot of troubles and seems a total mystery to the other side. "He's always so horny!" or "She's as cold as ice" are phrases we hear because of inherent biology. Men are very often able to "get in the mood" quickly, while women are often only in the mood once a month while they ovulate. And for either gender, when they are in the mood, they are IN THE MOOD. Watch out. "I look at a man and I get pregnant."

If you've ever said that, you might want to be careful and observe what time of the month it is when you even bother to look at other men. If you're struggling with a string of bad partners, it'd be a really good idea to look at your true motivation for picking your partners. Is it because of your own unconscious urges when you pick them? No way, you're saying to yourself. My intentions were pure.

Take care to remember that your biology and hormones are driving MOST of your decisions. Most of the time they're doing it right, and most of the time they are doing it without your input or awareness. Just like a flash flood that occurs on a sunny day from a thunderstorm 30 miles away, sex drives like to start innocently and gently, far, far away from the end goal, but they are still the motivator.

For the gentlemen, dudes, and guys reading this, have you ever laid in bed at night furious, fuming, or thinking non-stop about sex or sexual things until you feel like you're going to explode, or go crazy? Yeah, you gotta deal with it anyway if you want to live with others, keep your relationships, or your freedom. If you feel like you're going to lose control unless you do something, you're not alone. Dig deep, you can make it.

What's the worst you'll lose, as crazy as you think you'll go? Get through the storm and you'll be fine, better even. Don't give in. Your frustration, anger, and resentment will subside as long as you let that fire burn back down to the pilot light. And, yes, you'll survive, even if you don't feel like you will. Many women won't understand this, many will. It's your burden to bear, but you have to bear it. Sometimes you'll have a partner available to help you, but don't count on it and don't always expect it.

Sex is easily one of the biggest reasons for relationships to fall apart. Resentment and anger that builds over time because of unfulfilled needs is a rough problem for anyone to deal with. Frustration and resentment that builds over time because of the high expectations of the other person can be equally difficult and destructive. Talking about this, and learning to understand each other is crucial. Learning to sacrifice yourself, while the other person sacrifices themselves is what ties two unique individuals together. Imbalances will happen, things will be unfair at times. This is life.

As long as life and health are not in danger, keep trying! Staying together takes constant work and maintenance on both parts. Ultimately, the rewards keep us stable and functional, and our lives stay sustainable.

How does this all play into drugs and other addictions? Developing true intimacy while we deal with this crazy world and our unruly minds and bodies is extremely difficult. If you mix in any extra complications, it becomes near impossible.

Substance abuse can often start out as a result of innocent experimentation by kids and young people. Sometimes it's brought on by legitimate use of a prescription that ends up leaving an emotional or mental hole we need to be filled once we are off of it and we need to fill it with something else. How many people tried a drug, with the promise that it'll make them feel "good?" They'll "like it." Few of us have good enough reasons for avoiding something that seems innocent and will make us feel better.

Having a physical craving for a legitimate substance isn't a sign of an addiction, but it should be a warning light. There is a difference between "physical dependency" and "addiction". One is the use for a legitimate physical need, the other as a plug to fill in an emotional void. Perhaps a stressful situation or period of life leaves us looking for an escape, or a way to dull the pain of our surroundings. Most of these abused substances -opioids, narcotics, alcohol, whatever- have a way of dulling our senses. They turn down the volume of our frontal lobe which is shouting at us that something's wrong. Unfortunately, with the frontal lobe turned down, our basic, primitive, reptilian brain, that focuses more on satisfying immediate urges and drives, gets turned up.

After a while of listening to the outer part of our brain telling us that something IS wrong, or that we, ourselves, have DONE something wrong, we get tired of it. Like a nagging spouse or parent. "Just shut up already! Leave me alone." We muffle that noisy conscience inside our head with anything, just so that we can live with ourselves.

Often times there IS a real-life nagging spouse, parent, or friend that we are trying to escape or drown-out. Or some emotional scar we don't know how to heal, or even recognize is there. We can even become addicted to another person, or addicted to making another person better.

It's easy to attach our value or worth onto the actions of another person, usually a family member or spouse. We tell ourselves, "if my spouse is being dumb, it will reflect poorly on me" and so we nag and harp and manipulate and demean or intimidate the other person into behaving appropriately. Parents do this to children all the time: "If my child makes bad decisions, people will think I have been a bad parent." The sad truth is that these thoughts often turn us into the bad spouse or the bad parent that we are afraid of other people seeing.

By manipulating, nagging, coercing, guilting, or bullying our spouse or child into doing things our way, they often rebel in their own way and will go opposite our desires. Sometimes it's in overt outward rebellion, sometimes it's passive-aggressive and sneaky. The spouse or child slowly pulls themselves away and creates emotional distance from the situation to find their own safe-space away from the oppression.

And thus, in trying to reign them back in, we become even more controlling, mean, coercive, and insulting. We become co-dependent: our happiness is dependent on someone else's choices. And guess what the result of that is? More rebellion, dishonesty, lying, sneaking, angry, and reckless spouses and children, or employees, or inferiors, or soldiers, or whoever we are in charge of.

Rarely do addictions form in a vacuum all by themselves. Rarely is just one person to blame, either, though the addiction is still the addict's responsibility and no one else's. Rarely does a child or spouse do these things without trying to fill some void or try to escape their situation. Sometimes that void or poor environment is self-created by the spouse or child. Sometimes they made an initial bad choice and now they're trying to escape the consequences of it. Sometimes those consequences are perpetuated by loved ones. It's a really tricky landmine-filled realm.

A child that unknowingly tries an addictive substance simply because they didn't know better will retreat to that behavior or substance when life is hard or when they have to face consequences for their poor decisions. Sometimes children will reach out for addictive behaviors and substances because of overly aggressive, protective, or neglectful parents. Sometimes a spouse will become consumed by some destructive behavior, looking for the love they think they should be getting, but aren't. Maybe they really are getting it and are blind to it, or maybe they're not actually getting what they need. Each situation will be different.

The most important thing is to get rid of your own addictions first, whether it's co-dependency, drugs, porn, food, shopping, sports, TV dramas, or whatever is unique to you. Only when your own head is clear can you think logically and clearly enough to help someone else with theirs. ONLY then. This was illustrated in the Christian New Testament when Jesus described removing the beam of wood from your own eye before trying to remove a splinter from someone else's. This same idea is why airlines tell you to get your own oxygen mask on before helping others with theirs. You have to solve your serious problems before you can help others with theirs. Gandhi said, "Be the change you wish to see in the world."

I'd like to address a popular misconception about addictions:

There's the reports of how rampant Heroin use was among American soldiers in Vietnam. There were concerns about how many heroin addicts were going to come home and be let loose in the streets of America trying to blend their lives back into society. Yet, when these soldiers came home, there was no rampant heroin use.

A popular newspaper article was followed by a public speech by the same author touting that these soldiers were stressed and lonely in the Vietnamese jungle and once they integrated back into society among friends and family, their addictions simply went away. The answer to addiction is love and connection. When rats are given cocaine water, while alone in a cage, they will take the cocaine until it kills them. However, when placed in a cage with toys, distractions, and other rats, they won't touch the cocaine. He argues that humans are the same. All they need is love. There's truth to this, but it's also a quick way into a toxic relationship. "I just need to LOVE him more and he'll be okay!"

While this may be the case for some, in reality, a GREAT many soldiers who spent time in Vietnam turned to alcohol to calm their turmoil once they got home. Others did get into various other substances, but alcohol was the most popular, probably because it was the easiest to get. Data was collected and articles were written up in various professional journals showing that alcoholism among Vietnam veterans was rampant. They didn't return from Vietnam as Heroin addicts, but they quickly became Alcoholics.

Those soldiers didn't ask to be sent into a violent war, and many probably didn't know where else to get some peace and quiet other than the drugs, and many were simply never taught how to cope and deal with this kind of a situation, or didn't have the mental maturity or clarity to avoid the drugs while under that stress. No one should ever be condemned for having an addiction, but the condemnation should rest on whether the person is trying to overcome it, and what is their behavior like while they are acting on it.

We can't think —like this popular article and public speech claimed— that simply loving someone enough will get them to leave behind their addictions. It doesn't work that way. We all should learn to stand on our own two feet. It is a lot easier when surrounded by loved ones. A lone tree, left standing after the forest around it has been clear cut, is almost guaranteed to fall during the next storm, but that tree still has to stand all by itself even when it is surrounded by the safety of the forest.

Love and connection is critical to overcoming addiction, but it cannot be a substitute. The addiction does need to be addressed and handled as a separate issue from our relationships.

Some addictions are used to quiet our minds, others are used to turn it on. Because we're bored.

Stimulation is common in our society, with endless sources of entertainment, news, information, and thrills. I'm not about to say this is a bad thing. It's incredible! But we can get stuck on it if not careful. Endless hours on Facebook. Voyeuristic celebrity gossip magazines, non-stop streaming music, movies to no end, click-bait induced internet surfing. Complaining and negativity can become consuming because it feels so good to criticize others! The endless need for vindication, praise, and compliments, followers, and "Likes" can also become consuming and destructive. How many "Likes", "Thumbs Up" "Loves" and more can we get for our fantastic photo skills and artistic craftsy-ness?

How many of us know someone who argues, just to be contrary, someone who plays the Devil's Advocate ALL. THE. TIME. You know, because, we HAVE to offer the opposing view, all the time? Often, that stems from the need to look smart, or wise, by seeing things differently from the crowd. It's a need for recognition.

There's also the close cousin of mental stimulation: physical stimulation that goes for the physical endorphin rush of risky behavior. Adrenaline junkies. We've never had a time in our history where we can so quickly and so safely participate in what normally would be a dangerous, life-ending thrill. They're fun! And they can be consuming.

When we're doing physically-demanding, athletic things, we're learning and our brain is being stimulated and forming new connections, just as much as anyone studying and researching academic topics. There's a reason why collegiate athletes often find a great deal of success in their careers, even after they're no longer participating in their favorite sport. The brain is learning and growing whether we are stimulating it physically or mentally, and that brain growth will spill over into the other parts of our lives.

When we get hooked on stimulation of either kind there's not much difference in the end result: An overdevelopment and dependence on NEW things that appeal to our personality. Dungeons and Dragons or Fantasy Football. Comic-Cons or Super Bowl Games. They're both great and fun, but it's easy to get lost in both.

Now it sounds like I'm all preachy that the world around us and everything in it is bad. NOT AT ALL! Sorry if I did. None of these things are bad in themselves for what they are. They are only bad when we use them badly or when they are no longer constructive to life but destructive of it. And what you or I have a weakness for, others won't. I need to learn for myself what I am weak to and find appropriate ways to deal with it.

We need food and physical affection. We need stimulation. We need praise and respect. We need to be involved in the lives of other people. Participating in work, pastimes, and entertainment brings people together. Food brings people together! They all can uplift society.

Retreating and hiding behind an isolating addiction, or hanging out with friends in destructive behavior is the wrong path. It hampers true growth. We do it because growing wiser is hard, coasting along is way easy. We cannot hide from pain, awkwardness, and discomfort. We cannot avoid a bumpy life. We can't avoid others' unpleasant emotions. Feelings DO hurt, a lot! Living hurts. This world hurts. Interacting with anyone is going to hurt at times. I like what Dr. Robert Glover says, "They're just feelings, they're not gonna KILL you." Let's try to minimize it by learning the people skills we need to to minimize how much we hurt each other.

Not everyone will be weak to the same things. For one, alcohol or substance abuse will never even be a second thought. For others, it will be the crux of their life. For some, endless hours watching sports and sports commentators satisfy our craving. To others, it's video games or more scientific knowledge. The same can be said for others and their problems with "women" or food. Certain fetishes will appeal to some, and be appalling to others. Preferences will come and go as you grow older, or they may persist throughout life.

There is definitely a genetic component to all of these, but none of them need be a life-sentence that must be brutally endured, or simply given into. We get to decide who we will become, it's only a matter of learning and trying again and again. There are a great many negative things in this world that will push us toward the negative and destructive. We must learn what they are and deal with them appropriately. We can only grow food with care and knowledge. We can only stay healthy with effort. And we can only have loving relationships if we are working hard on it.

9

Why We Turn to Addictions and What makes them Soooo Good

I've already touched on this a bit, but I'd like to hit this point home a little more.

When I was in dental school, there was a mandatory substance-abuse lecture everyone had to attend. Dentists, with access to various abusable substances, are at risk and so the school organized an all-day seminar with guest speakers. Dentists, former dentists, a DEA agent, and a few others volunteered or were ordered by a judge to spend their time to come and share their experiences with substance abuse and recovery.

One guy said a statement that stuck out to me among all others, "Addiction is only the SYMPTOM of another underlying problem." That was revelatory to me at the time. The addiction is one problem, but until the germs of infection are removed, the abscess will stick around. Until you get rid of the fuel, the fire will keep burning. But then, even after the underlying void is filled, you'll still have to combat the persistent addictions and habits that will linger for a long time.

A popular acronym gets floated around various 12-step meetings, helping people remember the basic emotions that create addictions. BLAST: Bored, Lonely, Angry, Stressed, or Tired. I would add, Curious, Hungry and Horny to that. This is incredibly simplistic but gets the gist across. Many people would add another one: Being Afraid, or Fear. There are lots of other negative emotions. These are the Big 8.

So, what causes us to indulge in bad habits, escapes, self-medicating behavior, and bad habits? I already talked about sex, so here's the rest.

Exposure,Curiosity, and Peer Pressure

These are a huge part of addictions or aberrant behaviors. First of all, we won't usually want what we don't even know exists. This is why we protect our kids from things they can't understand. This is why we censor adulthood from our young families. We don't want to imprint or expose young, innocent minds to something that can leave a lasting impression that may cause them grief. Of course, we will need to expose our kids to the world, but only when it's age-appropriate. You don't need to teach a kid to hit for them to figure out how to do it. However, if they are exposed to it, they'll definitely do more of it, thinking it's normal or appropriate behavior.

Some aberrant behaviors are biological or instinctive. No one will need to show us what to do to get us acting poorly. Will we want everything we ARE exposed to? No, of course not. But there's a much higher chance of wanting to try it once we know it's there and become curious to try it out. Curiosity won't always kill the cat, but sometimes, it can and will. Curiosity often stems from Boredom. We don't usually feel curious to try new things if we're too busy with other tasks. Sometimes we become "curious" to find an escape from life if it becomes too stressful. Often, we use curiosity as a cover up for other negative emotions we haven't learned to recognize.

Few people are curious to try destructive things if there are no negative emotions pushing them to it. On the flip side, some of us are naturally-born explorers. This tends to be a very masculine trait, not just for males and men, but we see it a lot more in men. Many women are very surprised, disgusted, or mortified at what men are willing to try.

Randall Monroe draws the webcomic XKCD and has a brilliant comic called "The Difference." The drawing goes like this. At the bottom of the first set of panels it says, "A Normal Person" and it shows a stick figure looking at a machine with a lever. The lever gives a severe electric shock to the person. In the next panel, the Normal Person says, "I guess I shouldn't do that." Lesson learned.

The other picture shows the exact same scenario and is captioned, "Scientist." After the scientist gets shocked, he ponders, "I wonder if that happens every time" as they reach for the machine again. Some of us just need to figure out how everything works, and the downside is that we try to figure out how EVERYTHING works.

When we see others participating in an activity, and it, looks like they are having an enjoyable time doing it, some of us just have to try it. Or, we think we have to try it. We don't have a good enough reason in our head to avoid it or stay away. Like the scientist in the comic, who survives the experience, he wants to get a real grapple and handle on the situation and take charge of it. We do the same. If others survive an exciting-looking experience, we want to have the experience they seem to be enjoying.

The downside is that we often don't see the repercussions and fallout of people's mistakes, bad choices, or behaviors. Especially if they turn out to have been destructive or harmful. We hide it from the world, often out of embarrassment.

There's no shortage of exciting Youtube channels showing people doing extreme activities, extreme sports, or awesome accomplishments. What we don't see are the crashes, broken bones, destroyed relationships, hospital bills, chronic neck, back, or joint injuries that very often plague these participants through their lives. Like retired NFL football players who get to live big for a few years while they actively play, then spend the rest of their life paying for it with countless surgeries, physical therapy, or just permanent handicaps.

Boredom

Boredom, just like was mentioned in the previous chapter, comes from a lack of stimulation. An inability to stimulate your mind, or a lack of environmental stimulation. When people are bored they mindlessly browse the internet, often eventually settling into porn of some kind, or any kind. Or watch endless amounts of YouTube and Netflix. They might drink, by themselves, or eat a whole pack of Oreo's, or an entire Ben & Jerry's. Or go racing up canyon roads like they're some Japanese drift racer. They masturbate or indulge in fetishes over and over. They watch entire seasons of TV in one sitting. They go to parties where questionable substances are likely to be found and shared.

It takes a lot of conscious effort for many people to get their mind out of the autoplay habit that's formed in their life. Those habits usually form as teenagers and young adults, many times from seeds placed during childhood.

Water flows down the path of least resistance. Always downhill, always in some rut that has been carved out over time. Being NOT-bored requires a great deal of energy and thought and planning. Like pumping water uphill, it's an active thing that requires investment and forethought.

Sometimes, it's just fine and okay to be bored, to do NOTHING. In fact, a great many scientists, therapists, counselors, psychologists, life coaches, priests, rabbis, and others all say, you should set aside time daily, just for being bored! Let your mind wander. Let it go where needs, or meditate and try to quiet your mind by reducing all the stimulation. Maybe give yourself a primer, read something inspirational or thought-provoking and then go with it for 10, 20, 30 or 60 minutes. Meditate. Veg out. Relax. Practice doing nothing but thinking about your breathing in and out, in and out. Find a happy place. An isolation tank. Self-hypnosis. Take a nap, even!

I get that almost no one likes to actually be bored. I've seen a few studies showing that a decent portion of adults would rather experience pain than be bored. When assigned a time to sit quietly, with a device that delivers a small electric shock, a sizable portion of the people being tested will shock themselves repeatedly, rather than be bored.

Tell yourself, "It's OK to be bored!" A good thing even! Constructive! Letting your brain mellow out for a while is helpful and necessary! And not just during sleep, but during waking hours. Go sit on a park bench watching kids play in the snow or other people enjoying picnics in the sun! Or watch the clouds roll by and pick out shapes in them. Look for wildlife. These kinds of moments are where WISDOM comes from. This is the breeding ground for using your knowledge and facts and observations and putting them together in a way no one's ever thought of before, or at least you yourself. It's great to learn from others, it's even better when you can figure it out yourself in some quiet moment.

Cal Newport, in his book, "Deep Work" gives numerous examples of scientists, authors, and experts who accomplished their greatest work, or solved their hardest problems by dedicating hours each day, or months each year, to isolating themselves from the world, and be "bored" while they walk through the forest, or sit and think.

Galileo once said (as quoted in a worn out quote-book I used to have, once upon a time), "You cannot teach a man anything, you can only help him find it within himself." These quiet moments of boredom are those times when you teach yourself and find new things inside of yourself. Like bubbles of air in a mixed up frothy glass of water, letting things settle lets all of this stuff rise to the surface and get seen and felt and used. Treasures await you when you make a conscious decision to be bored and okay with it. And when you plan ahead so that you're not bored when you don't want to be, amazing things can happen. It's usually when boredom surprises us, or comes unexpectedly that gets us into trouble.

When you first decide to be bored, you'll find you can't get your brain to shut up! It'll just start spinning out of control. I listened to a Buddhist Monk describe it as "Monkey Brain" because it just chatters and chatters non-stop! But stick with it, find something to focus on —a painting, a scenic view, or your breathing— and with time, you'll find your mind getting quieter, more easily.

Loneliness

Loneliness comes in all sorts of forms. Simply being around people won't cure your loneliness. Often, when we're feeling lonely, if you can even recognize it as such, we're lonely for something...or someone, specific. Sometimes we're lonely for friendship. Sometimes we're lonely for companionship. Sometimes we're lonely for a superior, like a parent or grandparent, or mentor. Sometimes we're lonely for a guy friend or a girl-friend. Bros need each other, and Ladies need their nights out. Couples need their date nights. Families need their outings together and their meals together. And everyone needs someone to take instruction from and give instruction to. These are common and rooted in our DNA. By having all of these relationships, the entire human society is woven into this incredibly strong net and web. Like fabric. Let these connections go, and the net gets holes in it, the fabric gets weaker. A person becomes lost or at risk of becoming lost.

Why does society bounce back after a horrific natural disaster? It's these connections and the need to have these connections. We can mitigate many and most natural disasters by helping each other. We can soften the blow.

And when something does rip a hole in society, we can mend it with some time and effort.

Human nature doesn't want us to miss out on anything. We have such a strong need to be involved many of us get stuck with FOMO: the Fear Of Missing Out. We need to be part of the In-Crowd, the Cool Kids, the Popular group, High Society. Perhaps we can't stand any of those things so we rebel and go the other way, intentionally seeking the Outcasts, the Humble Folk, the Down-To-Earth. Either way, we feel the need to be involved with the people we look up to.

So, we glue ourselves to social media, to the news, to Twitter, and Facebook. The News. We need to be connected. We need to stay informed because we need to know what's coming to be safe. We need to have a side to belong to, a tribe, a clan, a nation, a community, and a family. There's a constant stream of criticism about our need for Social Media, but in truth, 100 years ago, people were addicted to Newspapers or Ticker Tape. Huge crowds, eager for any sort of stimulation, news, or entertainment would form around any sort of unusual event or happening out on a street, or town square, or other public meeting places. The addiction hasn't changed, just the form of it.

What happens when this drive to be connected and informed gets hijacked by too much information? When there's a need to be fulfilled, there will be people happy to fill it for us, and they want to and should be paid and compensated to do it. However, when our need to be connected becomes constant and insatiable, relevant and exciting news to feed to us runs dry often.

Really, we live in a pretty safe world nowadays. Even before modern times, armies usually took days or weeks to get anywhere and most people had some notice to run away or hunker down for defense. We like to think that when the British troops tried to sneak up on American Colonials, it was the bravery of Paul Revere rushing ahead on his horse who alerted the region of the coming attack. However, David Hackett Fischer's research has revealed at least 75 people who knew of this attack, at the same time or before Paul Revere did, and who spread the news within their own spheres of influence.

When we fuel this need for information and news nonstop, we become gluttons for news. We begin overeating and overindulging on it, gobbling any little bit of news we can find. The need to satisfy this craving becomes dangerous. We take some poorly worded statement from a politician and turn it into overwhelming proof that they are incompetent. We hear someone use a poor choice of words in a spontaneous conversation and we publicly crucify them over it. We learn a single statistic that can only apply to a small minority of the population or some small geographic region and we expand it in our heads to mean that we are all doomed.

Lately, there's been a non-stop stream of public figures having their distant past scoured and distorted out of context and then compared to our "enlightened" standards of today. CS Lewis called it Chronological Snobbery. This is when we look down on people of the past for being inferior to us today because we "know better."

We have news radio and talk show hosts and political commentators who can spend hours losing their temper at each other over three words spoken in some speech or award ceremony while we —as the viewers and readers— get sucked into it with righteous indignation. It all becomes this great voyeuristic circus, this endless entertaining coliseum of fighting and anger. The poor gladiators and slaves stuck in the arena of the public spotlight lose or win purely by public perception. We as the audience lose by getting so riled up in a buzz of adrenaline that keeps us slightly impaired for days at a time.

Who wins? Those starting the fights, those that are informing us. They get paid by their sponsors and advertisers for attracting millions of eyeballs emotionally glued to a screen like a moth to a bug zapper. When a terrible tragedy occurs, it is a windfall for every media outlet, knowing that they'll make big bucks on all the views they receive. There's little incentive for them to be truly helpful, except perhaps to put on an appearance of genuine concern every once in a while to have something to fall back on when critiqued about their motives. This applies to media outlets at all ends of the political spectrum, each offering to the public their own unique perspective of the events going on, to protect us from the skewed spin of "the other guys". The only real incentive is to keep us watching for as long as possible, to keep us occupied on the constant reporting.

Yes, help occurs because more people can be informed and that is good. More people can be involved because of how ready the world's information is. More people can help when they know what is going on. I don't want to condemn all news stations completely. An informed public is a strong public. We just need to be careful that what we're reporting and watching is based on a true desire to help and serve. Anyone who feels angered or scared by what's being reported needs to look carefully at the reporter and themselves, not always the event.

Mr. Rogers —from the 80's TV show, Mr. Roger's Neighborhood, "Won't you be my neighbor?" —often told of a time, as a kid, when he got anxious and worried about disasters and awful things happening on the news. His mom would tell him "Always look for the helpers." These helpers more often don't like to be on the news, they'd rather just do their thing, helping people who need it, not wanting fame or credit for it. Helping just because they love other humans.

Many people find their faith or belief in a loving God start to waiver and weaken when they see the horrible things we can do to each other, or that nature can do to us. They may wonder, "Where was God?" "Why didn't he stop this?" I don't know why God lets some bad things happen and not others. My kids don't always know why they get in trouble for some things and not others. But what strengthens my faith and hope in humanity, and what seemed to strengthen the hope of Mr. Rogers, is the ability to look for the "helpers." That is how we form relationships with people, we help them, and let them help us.

What if we don't have or never got to enjoy stable relationships growing up? What if our family was in shambles before we were born? What if we're stuck with Attachment Disorders? What if you just don't get along with other guys, or other women? What if you're single? That's where things can often go sour. In an attempt to fill one of these voids, we often entangle it with another. Maybe we didn't have a great role model for a mom or dad, so we mistreat our spouse or children, or —sometimes more sneaky— we overtreat them, we become codependent or "too nice" and afraid to rock the boat at our or their expense.

We might form romances with a good friend when we shouldn't. We treat our spouse like a buddy, a child, or worse, like a business partner, instead of an intimate partner. Watch out for these things! We become like those we have spent our time with. If you don't have a relationship you need, go look for one! In the kinds of places where you'll find the kind you need. Ask around. Sometimes a relationship you've already got can be changed into another, and sometimes it shouldn't.

An acquaintance can become a guide or mentor, but shouldn't be a master. An old friend can become a mate, or like a sister or brother, but not always an intimate one. A niece or nephew, or step-child can become like our own child. Work to become the half of the relationship you are needing for someone else! And don't stop trying, because in the trying we often find the fulfillment we are looking for. Just like the "real" relationships take work to maintain and grow, the alternatives require the same. If you're looking for a mother-figure, or a husband, or a child, you'll have to work at building the kind of trust those relationships are made out of, and you'll have to work at becoming the son, or wife, or parent that that person wants. We become like the people we spend our time with.

Kurt Yeager is a BMX cyclist who was injured and lost his leg in a terrible accident. Go watch the music video for Rudimental's "Waiting All Night" and you can see Kurt go through the long and difficult process of healing and sitting on the sidelines while all his friends keep biking and doing tricks. Eventually, eventually, he gets back on a bike and with help and tremendous perseverance he tries and tries until he's riding and doing tricks again.

The thought that occurred to me watching him sit by and watch everyone else doing what he loves is, "Why did he sit there watching everyone else and torture himself like that?" But I knew the answer. They were his group and by staying with them he could mentally prepare himself for doing it again himself. And it became the group that helped him do it, too!

When I was going through my separation and divorce, I knew I wasn't ready to give up on having kids or getting married. I spent as much time as I could with my still-married friends and their kids. It killed me inside, a little bit, to be around them and not be jealous of them. It hurt almost as bad as the divorce itself did. But, I knew it's what I needed to get back on my feet and not let my experience scar me any more than it had to. I wanted a family, so I spent as much time as I could around other families.

Don't abandon those you once trusted simply because you can't see their value to you, or your value to them, right now. Some addictions tend to isolate by nature. We think to ourselves that no one else will want to see us like this, or that no one can stand to be around us while we're binging. Others are afraid of showing "that side" of themselves because it doesn't mesh with who they want to be or would like to be, or who they think should fit into that society.

How many of us have a secret side to ourselves that we try to keep hidden away, deep down? How many of us would love to get rid of that secret side? The odd thing is that the secret side is such a deep core part of us, we are often afraid to get rid of it, afraid to bring it to light. Plus, indulging in that secret side is usually tons of fun, or at least, it feels really good to let it out once in a while. Like keeping a caged pet tiger. You gotta let it out once in a while for exercise, but you have to be careful where you do or you'll probably get the cops called on you. In a very real sense, this happened to a guy in Prince George, British Columbia, Canada several years ago while I was living nearby.

He had a permit to own and keep multiple tigers. He was traveling with them and felt bad keeping them cooped up in their containers, so outside his hotel, in the middle of the night, he let them out, 2 or 3 of them I think, so they could stretch their legs, get some exercise, and frolic a bit. At least, until some scared locals called animal control and the police! Imagine being in northern Canada and seeing a set of tigers running around in the middle of the night! I might call the cops, too. And maybe a therapist, because I probably wouldn't believe my eyes.

But pet tigers are extremely cumbersome pets and extremely dangerous animals regardless of how well they are tamed, and for the great majority of us, life would be simpler and easier and safer without them.

How often does our own secret "pet tiger" hidden deep, inside ourselves, cost us ridiculous amounts of money, or destroy relationships, or slowly kill the life inside of us and our bodies? Or worse, kill the life inside others? Like the real-life lions, tigers, and bears that some of us try to keep as pets, we never know when instinct will get triggered, and the wild animal inside of us will take over for even a brief period of time, causing havoc on our lives.

The stories of Werewolves, or the tale of Dr. Jekyll and Mr. Hyde play on this idea. We all contain a wild animal inside of ourselves that will not be controlled and only wants to selfishly destroy what's around us to satisfy our urges. It's usually unleashed by some trigger and doesn't stop until it wears itself out and falls asleep or the trigger is removed. Until we learn to avoid or properly address our own triggers we cannot harness that wildness for good, but become slaves to our DNA.

Other addictions thrive in the presence of others. We can't stand to be alone, but what do you do when all the people around us are participating in something that's fun and exciting and really bad for us? How many gallons of alcohol are consumed by college students at parties at the same time they're trying to amass the knowledge that's supposed to support themselves and others for the rest of their lives? You wouldn't want to know how many classmates would go spend an hour at the bar around the corner from the dental school instead of their dental classes.

How many kids get into drugs because of their friends? The great wide world is out there and some of us are going to get mauled by figurative bears and wolves and deal with starvation, hunger, and the worst of humanity before we "figure things out" with our friends, and some of us are going to get mauled by our friends before we figure things out and learn to leave them alone too. We'll pick up scars and baggage along the way.

Loneliness is dangerous, but so is being with too many people. We need to learn to accept both and live with both but in a way that keeps ourselves safe. Loneliness is not a bad thing and if we were to never experience or deal with loneliness we would never develop our own individuality. The opposite is true. If we are always hiding from others, we can never learn how to get along with, and live with, and help others, and let them help us. Biologically we are designed to be social animals, but too much sociality or not enough is harmful to us emotionally and mentally.

It's probably physically harmful to us, too. People who are married or involved in family environments tend to live longer, healthier lives. People who stay active by serving others in some capacity, whether it's paid or volunteer, find more purpose in their life and generally stay healthier longer.

How often do we fill our loneliness with pastimes and activities that we know just aren't that great for us, or that allow us to escape the harder task of being sociable with other people? We turn to video games, cats, books, or nature, and so many other things. Not that those things are bad, but if we are using them as an escape, our relationship with them becomes bad.

Being around people is hard work for a lot of us. It takes effort. It's tiring. And like any other skill, the more we practice it, the better we get. It's also easy to focus on the worst of people and justify our solitude. We have to be careful to not let everyone else's faults keep us trapped in our own small little world. We can be stronger than that and we need to be if we want to give anything back to the world instead of just taking and consuming: Consuming the video games that hundreds of people work hard to create, taking the pets that so many other people have bred and cared for to get them to us, or taking the knowledge and imagination of thousands of people who contribute to the books and stories we read or watch. The world only gets better as we get better and as we add more to it than we take.

Hunger

When we are hungry, we often can't think straight. When we are deprived of the most basic need of survival, we tend to get a bit crazy. In truth, whenever we are denied or blocked from getting any of Maslow's needs, our brain goes into overdrive stress mode to get what we are working so hard to achieve. However, Self-Actualization isn't quite so necessary for survival as Hunger is. When we are denied food, our brain will become fixated on doing whatever it can to find it, or some substitute for it. Or, something that will take our mind off of our hunger, which can be almost anything that is stimulating enough. Easy to say is that hunger causes us to not think normally, until we learn to control it, or learn that food will be coming soon. Like I said earlier, we become Hangry. Ask any mom who is late feeding her baby or a toddler, or even older children.

Yet, as science shows us over and over again, learning to enjoy the feeling of hunger, or "fasting" from food for short periods of time, is very healthy for us, mentally, physically, emotionally, and even spiritually.

Anger

This tends to be the most outwardly destructive and visible of these negative emotions. It's hard to hide the fact that you're yelling and chewing somebody out. Or throwing chairs across the room. Or stomping down the hall. Or dropping the f-bomb every other word just to show everyone in ear-shot that you mean business. Or constant complaining, nagging, manipulating, insulting, and on and on.

Anger stirs up our adrenaline, which stresses us out, makes us afraid, uncertain, and often scared or worse. It hurts us to be angry, and it hurts everyone we target. Kids that are the target of anger will develop symptoms of abuse when exposed to it for long periods of time. Anger can twist and distort our perception of reality in the same way that alcohol and drugs can. Anger blinds us to a great many details of the world around us. All of the negative emotions in this book can do the same, but anger gets the most attention for it. All that extra attention may not be entirely justified, since loneliness and boredom blind us to a good part of the world around us, too, but since loneliness and boredom tend to be emotions of inaction while anger is an emotion of action, it's easy to spot.

When we are angry, we tend to say things we normally wouldn't or don't even mean or believe. Think of all the relationships ended during a fight, and regretted later. Or the mean things said to loved ones that linger for years. Or the outbursts with strangers that turned into legal problems.

That's not to say that all yelling, complaining, or nagging is out of anger. I can yell at my kids and not be angry with them. Sometimes it's the only thing that'll get their attention. Sometimes they just need to learn really fast to put down the knife while they run through the house!

I've heard people say that anger is the opposite of love. It's not. We wouldn't be angry at people we didn't love or didn't want to love. Even strangers. We care about our country enough that we get in fights and say dumb things to fellow citizens who disagree with us on politics. And we get angry at them because we are afraid that their philosophies and ideals will affect our happiness. Anger is somewhere in between love and fear, separate from each, and it can be motivated by both at the same time.

Some will tell you that there is "righteous" anger or constructive anger. There is, but it's rare and only works as a very last resort, and often, doesn't work, even then. In the vast majority of circumstances, it's best to avoid anger, or lecturing, or belittling. When I've lost my patience during an internet argument, it makes me look like an idiot, no matter how intelligent I think my facts are. When angry mobs incite riots on the evening news across the country, burning down local businesses, hurting innocent people, or protesting the very groups that make their livelihood possible, they are often proving their own stereotypes. Anger tends to reinforce whatever negative tendencies we already have and show it to the world. Violence, victimization, manipulation, coercement, hypocrisy, incompetence, laziness, etc. It magnifies our flaws.

Anger is best used as a passion that drives us to constructive action. When we are frustrated, we can channel that frustration into work and effort to pull ourselves out of a rut, out of a stressful environment, or out of the doldrums. I've been known to do weird things when I'm angry, like fix a long-broken project, or mow the overgrown lawn, or buy my wife a deep freezer. Yes, I bought my wife a deep freezer when I was mad at her. Those are sometimes the only things I can think of doing that doesn't involve yelling and losing my temper. When I read this part to my wife, she asked, "What were you even mad at me about?" I don't remember. I don't have a clue. I just knew I needed to do something nice for her to remind myself how much and why I still love her. It worked.

It's no coincidence that Alpha, "red"-type personalities are associated with being hot-headed and having strong tempers, and are also known to be doers, achievers, and the strong personalities that accomplish a great many things for good.

For most of us, Anger is an unpleasant emotion that we don't like seeing inside of ourselves. We feel out of control when we experience it, and that is scary. Or the stress of whatever is triggering our anger is taxing on our minds and our bodies. It can be crippling and disabling over time. It can get our blood pumping and our stress hormones cranking and those effects can leave us recuperating for days or even weeks.

Anger can be an addiction in itself. It is a difficult emotion to control and stay in control of and with it being an emotion of Action, the "rewards" we receive while we angrily get our way can be difficult to turn down in the future. When we get our way, we keep doing what it is that gets our way. Think of a spoiled child who always gets their way or a tyrant boss who makes lots of money at the expense of their employees.

Practice, practice, and more practice can turn that fire into something useful when contained. If you touch a spark to a gallon of gasoline you get disaster, but when it's properly contained inside the right kind of engine, you can move 2000 lbs of machinery really fast from one city to the next. Vroom.

Controlled passion and anger need a steering wheel to get where you need to go. That steering wheel is our intelligence, our knowledge, our wisdom. A pre-frontal cortex of our brain. This doesn't mean we don't steer with our emotions, it means we steer with our conscience.

Anger is stirred by our most primitive parts of our brain, and it usually goes in one of two directions: Fight or Flight. Silence or Violence. Many of us retreat in our anger into our own lonely solitude, or by spewing that anger to other people who we think will vindicate us. Others get angry right back at the source of their anger, fighting back until they beat down whatever it is that threatens them. These are the natural reactions that our primitive brain thinks of.

But what if there were a third way? Door number three. Somewhere where you got the point across directly, but without the insulting, anger, or victimization that most of us resort to when we feel resentment and anger. I'll cover this more a little later, in part 2 of this book.

Stress

Stress is the response our body goes into when it's threatened by....just about anything. Sitting on top of your kidneys, like some king on a throne, are your adrenal glands. They, like henchmen in the employment of your brain, can flood your body with adrenaline and other hormones to make your blood vessels open up to get your body moving faster, your eyes dilate to pick up more light, your brain to activate and process information faster than usual, your muscles to fill with creatine and sugar for fast action, and your digestive system to slow down, so that you can spend all your energy getting out of, or dealing with the danger or threat.

Not just hormonal, there is a nervous, electrical side to stress. Popularly known as the GO and KNOW systems, the Fight or Flight vs Rest and Digest systems. The Sympathetic and Parasympathetic systems. One floods your body with monoamines —adrenalines— the other slows your body down with Acetylcholines and the natural "nicotines" I mentioned earlier. Cytokines are like roving molecules out looking for problems to fix and when we get too many of them, they don't always fix things correctly or try to fix things that don't need fixing.

Along with the hormones from the adrenal glands like cortisol and adrenaline, and the hormones from your brain, it shuts down anything that's not important for the immediate moment. Things like your stomach, your sex drive, your hunger, your intestines, your need for sleep, the healing of wounds and fighting off infections. It ramps up your muscles, your brain, your heart beat, and it can do it real fast. It's the gas pedal and ejection seat of life.

Whenever you get stressed, you will find some of your muscles tensing up, your breathing getting faster or shallower, and your heart beating a bit harder. It's hard to directly control your heart, but you can control your breathing, and most of the time your muscle tension. Slow them down and relax them as best as you can. Often your heart will follow the others.

The alternative to the stress-system is the "parasympathetic" nervous system. It slows down your muscles, it calms your brain, it floods your body with oxytocin and dopamine to feel good, it pushes food through your stomach, it makes you feel safe, it perks up your sex drive. It nudges you off to sleep and gets all the repairs and rebuilding of your body underway.

As a side note to this, the sex drive does some weird things. Both the stress and relaxation systems are heavily involved and used during sex. Both parasympathetic and sympathetic systems light up like crazy. Many people report being extra "horny" when they are sick, while many people say its the last thing on their mind. Sometimes when we are filled with rage, stress, and anxiety our sex drive runs rampant with it. It can fire up as a last ditch effort to pass on your DNA before death sets in or to calm down and soothe our fears. It can shut down during danger as a way to preserve your own life. It will turn on or off depending on a whole lot of things.

There's been gobs of papers, books, stories, documentaries, articles, journals, newspapers, and blogs devoted to stress. It's a major issue in our culture. We are driven to go and do, to work and accomplish, and try, try, try until we are successful. It's a contributing factor to many of our serious illnesses, including: Heart Disease, Auto-immune disorders, Fibromyalgia, Arthritis, Alzheimer's Disease, and more.

Stress keeps us functioning when we are in danger, but the downside to this is the toll it takes on our body and our minds. Our immune systems suffer and we end up with all sorts of chronic diseases and injuries that never heal. Have you ever noticed how you'll get sick after some long-awaited stressful event is over? How many students will get sick and their bodies will crash right after a Final Exams week, or a semester ends? The reason? Your body is keeping you going, worried about the stress of it all, and once the thing is over, your body can finally take a break, slow down, and deal with all the backlog of repairs, infections, and trauma that it's been ignoring for the past several weeks. It shuts down for major cleaning.

When we've been through a traumatic experience, our brain suffers and we are stuck with memories of the traumatic events that never get sorted out, played-through, or resolved. Your brain needs to deal with those awful memories and experiences and play them through all the way to the end before it can package them up and put them away somewhere safe. Like sitting down to watch a movie ten times but you always get scared and don't finish it. So you never find out what happened. Our memories often need to do the same thing. Many times we have to "watch" those memories over and over until they no longer sting or hurt us or make us flinch when the scary parts come on. Without that resolution and closure, we begin to slowly treat everything with mistrust and fear. We can't always get "closure" from the thing that caused it, so don't expect it. Instead, we may just have to look for realistic understanding and close the case ourselves.

Anxiety and stress can become center-stage of our thoughts for so long that we can't turn it off or get it out of there. It takes over all of our judgments, thought patterns, rationalizations, and decision making. And we often become unhappy through all of it. Depression or PTSD start to run rampant. Some of these memories of stressful times can take over our whole body and we lose control of ourselves, distorting our reflexes, making us afraid of normal things. We need some sort of safe place where we can play these stressors out until they lose all power over us.

When stress dominates, no longer can you see the positives and negatives of optimism and pessimism, but pessimism starts to seem like the only thing that truly exists. Stress, along with Fear, becomes the filter you see life through and it can work backward to re-write all of your past memories with its tint. Optimists get accused of seeing the world through rose-colored glasses, but pessimists see the world through dark-colored glasses. The point is that everyone tints the world to their color of choice, some of us tint it blacker than others and usually not on purpose or even with our awareness.

Our body and our brain need some sort of "out" to escape all this stress and anxiety. It will look for distractions if we don't give it appropriate ones. It will look for sedatives to force some rest on yourself. Unfortunately, many of these sedatives and distractions only force our body to work harder in the long run in exchange for a quick flood of dopamine, oxytocin, serotonin, or GABA, and the stress cycle gets worse and worse and we need more and more of the drug, or harder and stronger versions of it to keep escaping the pain.

Your body is looking for a coping mechanism. A grip on reality. A way to process what is happening to you. And nature did not give us the ability to do that by ourselves on our own. Nature, or God, or the Universe, or whatever you want to call it, programmed us to get together with other people to resolve our problems, and in return, help them solve theirs. We talk to each other, we encourage each other, we help each other, we support each other. We love each other, and thus we get a HEALTHY dose of dopamine and oxytocin.

Our brain processes the stressors at least twice when we work with others to resolve them. First, in forming concrete thoughts about them and then second, to speak it out loud. Additionally, and thirdly, we get the knowledge we are lacking and some additional brain power from others who can process the problem in their own way and hand it back to us. Our brains were not organized to be independent all the time.

Slowing down, getting a good night's sleep or a nap, eating a healthy meal with friends and family, in a calm and safe atmosphere is one of the best things we can do to reduce stress. Control your breathing. Take charge of yourself by slowing down, intentionally. Quiet time, meditation, prayer, isolation, are all important, too, and this will come up again later in this book. Deciding to be "bored" or just doing nothing can help, too. Various other therapies are helpful and often needed, too.

In many ways, the opposite of Loneliness, stress is often induced by too much time around others. There's a pattern to some people who've lived busy, often very successful lives in the cities of the world, who then retreat into the country and woods to escape "it all" for their retirements.

Tiredness and Fatigue

Similar to stress. When we are tired, out of energy, or worn out, our brain needs rest and a chance to process the previous day and the latest stressors and memories, to clear itself of excess baggage. It's like a clogged heating and air conditioning vent. Yes, air may still move through, and because you're so used to how little comes out, you forgot what it COULD be like if the vent were clean.

Like playing outside in the dusk. You gradually become accustomed to the darkening sky so slowly you can't realize how poor your vision has become.

Being tired and fatigued inhibits our brain from thinking clearly. Similar to being drunk, our ability to make rational decisions is gone, whether we think it is or not.

A study was done by the New York Times in 2015, using Google's data on internet searches about the types of internet searches done at various times of the night. A little before midnight, searches for online dating sites reaches its peak. Not long after midnight, terms related to pornography hit their peak. And about an hour after that, between 1:30 am and 3:00 am, searches related to "loneliness" and "suicide" hit their peak. This shouldn't surprise anyone who has gotten this far into this book. Also interestingly, the deep philosophical questions about "the purpose of life" or "the universe" are most searched in the middle of the night, right alongside questions like, "How to roll a joint?" which also peaked between 1 and 2 am.

Our brain doesn't actually do it's the best thinking when it's tired. Despite what we like to tell ourselves, we are not "all there" when we're sleep deprived. Important parts of our brain just aren't working well, while other reckless parts of our brain are working without restraint. While it's tempting for me to write this book late at night when there are fewer distractions, I've learned enough about myself to know that my writing gets really sloppy. I'll go back and re-read what I wrote during some late-night session and have to stare at it, thinking, "What did I actually MEAN there?" I'll skip entire ideas, words, and sentences. If parts of this book don't make sense to you, chances are, I was over-tired while writing it, or you're overtired while reading it...or perhaps... I just need to get better at writing.

The middle of the night is also when medical symptoms are most searched. In other words, people are looking for answers to their problems. Our brain wants to do a lot of problem-solving at night. It's designed to do that while we sleep. If we're not asleep when we're supposed to be, it'll drive you to do what your unconscious was designed to do, process, fantasize, and dream, except that you're awake and getting drugged by whatever it is you're doing. When we get our fix and high of whatever it was we needed, we have to come down off of that unhealthy dose that came from unnatural sources and we crash into despair and depression. Our mind, finally worn out, loses it. Multiply this pattern a few nights per week for several years and we'll start to see major problems crop up in our lives where few used to be. Multiply this over millions of handheld devices and laptops and PCs across the country, continents, and world, and it's easy to see where large societal problems can crop up.

In a few recent studies in the past couple of years, sleep problems are being connected to diseases like Alzheimer's and Parkinson's disease. A lack of sleep builds up normal proteins between our brain cells and as these proteins clump together during our sleepless nights, it begins to affect our brain tissue and how well our brain works.

Ohio State recently published a study where they measured the amount of sleep a couple got, how well or poorly they resolved a conflict in their marriage, and also how stressed their bodies got because of the conflict. With less sleep, these couples would get into more hostile and aggressive arguments, AND the amount of stress and inflammation their body went through would spike because of it. Moral of the story? Get sleep so you and your partner don't argue as violently, and so your body doesn't suffer either. A good idea is to set an appointment with your partner to discuss the problem at a better time when you're both well-rested and can safely talk to each other. It's no secret that when we don't get enough sleep we are irritable, cranky, tired, distressed, and more violent.

We need sleep. A decent amount of it. Men TEND to need a LITTLE bit less than women, on average, while teenagers and kids need a lot more than they typically get. In the end, we all need at least 6-9 hours a night. This gets really tricky for people who work night shifts or weird hours. Often-times our "fear of missing out" (FOMO) gets in the way of getting the rest we need. We don't want to be asleep while parties are going on, movies and shows are playing, or while other people are just awake and enjoying each other's company. Sometimes we just don't want to put down the game controller or Netflix, just to get some sleep, because, why quit on the fun stuff?? Ugggghh, sleep is such a CHORE!

My own mind is so intent on learning and being stimulated, it is hard for me to tell myself, "Enough! Time for bed." I'm grateful for a wife who likes to sleep. Going to bed at the same time as she does is extremely helpful for me. And by letting her dictate some of our bedtimes I can overcome a tendency I've dealt and struggled with my whole life. When I was single, I got sleep, sure, but a lot less of it, unless I was really careful with my schedule and night-time rituals shortly before bed-time.

Each person will have to form a routine. A ritual. A pattern or habit that will get them to bed. Start that ritual around the same time every night possible. Even if it's an hour-long or two-hour-long ritual, go for it. Dogs and cats have the curious ritual of walking a circle or two to pat down the ground beneath them before laying down to rest. We humans have equally curious rituals that are incredibly more complicated. Use that to your advantage! Form the pattern when you're wide awake and sober, not when you're already half-tired and groggy.

Make the decision when your head is clear and begin the ritual while it is still clear. Brushing your teeth, watching a show, snuggling with a loved one, drinking a cup of tea, taking a shower, washing your feet, whatever it is, —whatever you need it to be— practice it. Plan it out, and let the autopilot you consciously arranged for do all the work. My wife's ritual takes her 20 minutes. Mine takes about 3. And I've learned to time my ritual to hers most nights so we're ready for bed at the same time.

What if you can't sleep? Oddly enough, for many people, the biggest killer of a good night's sleep is worrying about not getting enough rest! The anxiety we have about being awake does more to keep us awake than a lot of things. A number of health problems can ruin our sleep. I don't want to minimize those issues. I hope to instill in everyone the notion that worrying about your sleep is the worst thing you can do. Worrying about anything is the worst thing you can do. Don't think you're the only one kept awake by the worst of your thoughts and fears. Your Demons. Everyone will deal with this throughout their life. Learn what you need to, to understand that any rest is beneficial, and often it will be enough to get you through the next day.

Even if you're not sleeping, laying in bed and resting is still resting, and it's incredibly valuable. Do your best to enjoy it. Your mind is going to work away and churn as it normally would while you were asleep. It's frustrating for most of us when it happens. Learn to embrace it, and let your conscious thought step by and watch your brain work, like a spectator. You may wish for better progress to be made as you cycle through the same problem over and over, or the same song lyric on endless repeat. That's okay. Just trust your brain, and trust your day-time sober mind, more than your tired and mixed up mind.

Often, your brain is stuck awake, overflowing with unmet emotional needs, or pushing like a madman to relieve some pent-up and overpressurized hormones (sexual or other), needing an outlet. Maybe you're dealing with huge insecurities about yourself or your relationships (or lack of them). Maybe you're afraid of an important work meeting. Maybe your brain is still processing that deeply profound, or traumatic movie you watched earlier.

If you need relief from some pent up sexual pressure, your brain will fantasize about all the wonderful ways it could get that relief. Often, all you need is to find some guilt-free way to relieve that pressure temporarily and those fantasies will subside or turn down hugely. It doesn't mean your brain will quit, but you can at least turn down the volume on those thoughts.

In a worst case scenario, you don't have access to a guilt-free way to relieve that pressure (while many of you will feel totally comfortable with masturbation, many others are not), and you'll endure a long night of uncomfortable stress. It will feel like a major endurance event. Remember that you will survive. It will be okay. You will not lose anything over enduring it. One of two things will happen. 1. Your hormones will get so backed up they will revolt against your brain and organs, demanding they slow down production, or 2. You will realize that you can survive those intense moments, however long they seem to last, and you will gain greater confidence in yourself for it.

You are not your best self when it is late. No one is. If your partner or spouse is insistent on solving your problems at 1:00 am, be insistent that you take some time to cool down, collect your thoughts, and make an appointment during the daytime, sometime, to discuss it. Do whatever it takes. If you want to become a better person and help the world be a better place, start by just getting enough rest.

In his book "Deep Work," Cal Newport describes the ability to set appointments for yourself and your partner to tackle problems later, as critical for being able to calm your mind and focus on more important tasks. Take full advantage of it.

Fear

Fear is the grand-daddy of all problems. The ultimate enemy. The Adversary. Satan. Entropy. Chaos. Whatever you want to call it. Like I said earlier, some people think anger or hatred is the opposite of love. Anger is the collision of fear and love. Fear is the opposite of love. Fear is the source of all our problems in the world.

Fear causes Stress. Fear makes us worried. Fear keeps us awake at night worrying about the future. Fear tells us that Anger needs to be aggressive or we'll be treated like a doormat. Fear tells us to hit first before anyone hits us. Fear tells us to leave the relationship before the other person does. Fear tells us that loneliness is the worst thing in life (when it isn't), or that our loneliness will be permanent (when it won't). Fear of doing and failing, and accomplishing and working leads us into laziness and boredom. And worst of all, fear spreads like cancer. It spreads within ourselves and it can spread to others.

Fear tells us that our problems cannot be overcome. Fear tells us that others will make fun of us. Fear makes us embarrassed. Fear turns our normally appropriate shame into anger and defensiveness at others for calling us out on it. Fear keeps us silent when we need help. Fear can cripple us.

Fear of starving or being poor is what leads us to steal. Fear of not getting sex is what pushes us into looking for its substitutes. Fear of being outcast is what causes us to cast out others.

Fear is darkness pretending to be light. Fear is ignorance imitating knowledge. Someone who overextended themselves in some assertion of "fact" will be afraid of being caught wrong and create numerous problems in the world around them for fear of being "wrong". How much trouble to little children cause trying to cover up their little fibs? How much MORE trouble do world leaders and politicians cause, trying to cover up their mistakes?

Fear is mistrust. Fear is dishonest. Fear is fakery and deception of the ultimate kind. Fear isolates us, chews us up, spits us out, and leaves us in the dark —abused, broken, and crippled. Fear makes us, Irrationally Insecure. We feel so vulnerable that we become irrational, unreasonable, crazy, selfish and monstrous. Often, Fear is driven by hormones, whether they are cyclical or not. Women often feel Irrationally Insecure near the beginning of their monthly periods. Men often feel Irrationally Insecure when they don't know when their next romance is going to come. We are not our usual selves in those moments.

Wise are the words, "There is nothing to fear but fear itself." stated by

Franklin D Roosevelt during his inaugural address and adapted from Sir Francis Bacon.

Yann Martel, in "The Life of Pi" expresses this wonderful thought about fear, "I must say a word about fear. It is life's only true opponent. Only fear can defeat life. It is a clever, treacherous adversary, how well I know. It has no decency, respects no law or convention, shows no mercy. It goes for your weakest spot, which it finds with unnerving ease. It begins in your mind ... so you must fight hard to express it. You must fight hard to shine the light of words upon it. Because if you don't, if your fear becomes a wordless darkness that you avoid, perhaps even manage to forget, you open yourself to further attacks of fear because you never truly fought the opponent who defeated you."

I got this one from a favorite kid's show, "The important thing is to talk about our fears because if we don't, they can throw us out of balance...Admitting your fears is the first and most difficult step in overcoming them." (yes, that is Nickelodeon's "The Legend of Korra")

So how do we grapple our fears? How do we overcome this true opponent? How do we put words and shape to something that is indescribable and shapeless? How do we beat that part of ourselves that wants to fail over and over and over again? How do we get control of ourselves and take back what is rightfully ours, that right and privilege of self-determination that belongs to every human being just by being born?

First, name it. Give it an identity and shape. Describe it, in as much detail as possible. This puts a container on it. It gives it boundaries and limits. I gave some examples at the beginning of this section. Next, practice. Little steps at a time. Get up. Again and again. Your mind is strong. Stronger than fear. Stronger than you will ever realize. Love for yourself and love for others are the reinforcements you need to overcome it. Love for some goal that is bigger than yourself. Realize that as long as you are alive and thinking, you have the ability to create, to be creative, to solve problems, to overcome. Even if you can't overcome the problems that hurt you most, the act and practice of overcoming other problems will bring you the joy that will eventually defeat all fear in your mind, whether you make a full recovery or not.

Yes, we all will have lifelong disabilities of some form or another, many that were beyond our ability to ever control, some that were blatantly and foolishly our fault. But if we are alive, we can improve. We can get better, and we can make something of ourselves that will inspire others.

Nelson Mandela was imprisoned for 27 years for trying to improve democracy in South Africa. Not long after his release from prison in 1990, he was elected as President of South Africa in 1993 in a triumphant and extremely inspiring story. In his inauguration address, he used this quote from Marianne Williamson's book "A Return to Love":

"Our deepest fear is not that we are inadequate. Our deepest fear is that we are powerful beyond measure. It is our light, not our darkness that most frightens us. We ask ourselves, Who am I to be brilliant, gorgeous, talented, fabulous? Actually, who are you *not* to be? You are a child of God. Your playing small does not serve the world. There is nothing enlightened about shrinking so that other people won't feel insecure around you. We are all meant to shine, as children do. We were born to make manifest the glory of God that is within us. It's not just in some of us; it's in everyone. And as we let our own light shine, we unconsciously give other people permission to do the same. As we are liberated from our own fear, our presence automatically liberates others."

I want to emphasize the need to label our fears, to describe them. We need to put a "handle" on each and every one so that we have power over them, instead of letting it have power of us.

All the best horror movies scare us by hiding things in the darkness of an empty house, in the shadows of the woods, or a cave, under deep waters, or out in the dark emptiness of space. All those places filled with "unknown" creatures. We are afraid of what makes us uncertain, vulnerable, exposed, or when we just don't know what's out there.

We teach kids to not be afraid of their closet, or under their bed, or their bedroom by turning the lights on and showing them what causes those scary shadows.

The 2017 release of the Stephen King movie, "It," seemed to be doing well enough, but among the critiques of the movie that were repeated were that the creature was not as scary as moviegoers feel it should be. The audience got to see the clown-monster Pennywise early and often, and they lost their fear of what they could plainly see.

Many people think it's foolish and silly how certain religions try to personify evil. To teach about a Devil like Satan (the word literally translates to "Opposer"), a darkness, or Void, Mara, Djinns, Ahriman, or the various gods of Norse mythology and other world religions, is to put a name to our fears, or to explain away the scary and uncontrollable things we see. Science teaches of Entropy, that everything will eventually descend into disorder and into cold, lifeless darkness.

Knowing what we know about how dangerous fear is when it's a nebulous shadow force, without boundaries or physicality, that we cannot control or manipulate, it seems like a wise and prudent thing to put a name to it. To put a physical form to it that we can describe and see and manipulate and protect ourselves from. To learn of its habits, its tendencies, its patterns, and then stay one step ahead of it. When we personify evil or fear, it becomes something we can avoid, or fight, or keep away from us and our loved ones.

And what is it that we are afraid of? We are mostly afraid of things that are just a little different from us. We are afraid of change. We are afraid of things that disrupt our lives, our environments, or the status quo. From little children being instinctively afraid of grown men, to politicians being afraid of an informed public. From those with different colored skin, to those from a different background or raised in a nearby region. There's a strange amount of animosity we all feel to those closest to us. The sibling just older or younger than us. The city or county just over the hill. It's easy to think that the people in the nearly identical community a few miles away is crucially different. They are snobs, or rednecks, but in reality, they're almost the same as us. It might be the country that shares our border or is closest in ethnicity and culture to us, such as the USA and Canada, or Japan and South Korea. Or, the other Christian church down the street, or the other Muslim sects or tribes. We often feel most threatened by our neighbors, and less so by those farther away.

Let's use the very loaded and heated topic of Racism to illustrate this:

We are naturally afraid of what we think is different from us: skin color, eye color, hair color, language, dialect, glasses, clothing, lifestyle, traditions, hand-preference, etc. We won't all see or worry about the same kinds of differences as everyone else, and we won't all put as much weight into those differences. Some of it is taught, some of it is natural. Some of it is due experience. Some of us don't care about hair color, some will. Some won't care about styles of talking or preferences of music, some will.

If members of a group of people have treated us a certain way, our brain will make the automatic shortcut that it must be a group "thing" to be like that. Our brain does that with every aspect of our lives. It saves on decision-making brain-power so we can worry about other new things we'll have to deal with. Once we've figured out our daily route to work or school, our brain will go into autopilot taking us there, without us consciously having to think about turning left or right here or there to get us where we need to go. It saves energy and our brain does it automatically.

When we've developed a ritual for going to bed, it becomes automatic. When we've developed the patterns that help us shoot a basketball or swing a baseball bat, it becomes automatic so we can focus our attention on other details, and thus our skill increases and we become better at the task because we don't have to consciously decide every little thing anymore. While writing this book, if there's a section I know I need to put more thought into, I turn the text a different color other than black, so it disrupts the flow, and I have to look at it and read it more carefully.

So when we encounter someone from that "group" that has scared us in the past (whether rightly or not), it's hard to let go of preconceived notions, prejudices, and assumptions. Those are natural. BUT! we all must allow our brain to make exceptions when we find them and add them to our pool of knowledge. When we've met enough people from enough different groups, two things happen. 1. We will have lots of prejudices and we will have formed a lot of stereotypes, and 2. We will allow for lots of exceptions to those prejudices and stereotypes.

The more I understand this world, the more I can predict about it, AND the more I am surprised by it. I begin to understand the patterns of different cultures, different ethnicities, different countries. I also understand that "people" are "people" everywhere, and the same kinds of problems and strengths of one group of people will be similar or the same as those of just about all groups of people. They may look different because of traditions, clothing, words, or appearance, but deep inside, the problems and advantages to humanity are common to all of us.

So don't be angry at each other for having weaknesses! We just need to teach and inform each other, —lovingly— as best as we can. There's a popular phrase that "racism" isn't born with us, we have to be taught to be racist. That's only partly true. Some of us will ignore skin color, some of us won't. Some little children will naturally be afraid of another child that looks different. Some won't. And sometimes you'll see different results within the same family. Whether you are a white person living in Japan, or black person in America, or even between subgroups of ethnicities of the country of India, or between the lighter and darker skinned groups of people in the Caribbean or South American countries. Racism is everywhere. It has always been here and probably always will be here, but as we get better teaching each other about it, we can learn to compensate to a large degree. We all profile each other and we will all be profiled by others. Sometimes to our benefit, sometimes to our detriment.

Most times, our pre-formed assumptions and snap judgements don't affect us very much or can make for some interesting stories. One of the funniest examples of this I can think of happened to me when I was about 14 or 15 years old. I was a slightly overweight, round-faced, glasses-wearing white kid in suburban Camarillo, California. I was working for my dad's small health food grocery store. I was not my full height and probably only around 5 foot 6 inches. I was wearing a long work apron with the store's logo on it. My hair was probably evenly combed and parted nicely to the side. I was definitely not the coolest kid you ever met.

I walked out of the store to retrieve shopping carts from the cart-return corral. A pair of women had just parked next to it and were getting out of their car. The younger of the two, probably in her forties, eyed me suspiciously as I walked up to the corral. As I got close she stiffened up. As I got closer, she clutched her purse to her chest and squinted her eyes at me. I thought she was acting a little peculiar but just did my job. As I turned into the corral to grab the carts, she visibly relaxed, let out a big puff of air and laughed. "Oh dear!" she says, "I thought you were after my PURSE!" I just smiled an awkward smile, gave a courtesy laugh, and kept doing my job.

For whatever reason, she was scared of a teenage boy walking towards her, despite so many other indicators that I was probably not going to harm her. Her fear got the best of her and she could only focus on defending herself and protecting her belongings. I can only imagine what her previous experiences with teenage boys had been that she would be so scared of me in such a safe set of circumstances and me looking like such an innocent nerd. Really, if you could see pictures of me as a young teenager, I expect very few people would be intimidated by me, and yet she still was very intimidated by me all the same.

Unfortunately, stereotypes can be much more harmful and destructive. Sometimes leading even to unnecessarily taking others' lives over it. Humans have killed each other over no other reason than a person's nation of origin, ethnicity, skin color, religion, last name, neighborhood, clothing, and attitude, or just being in the wrong place at the wrong time. We learn through our life experiences that people who enjoy similar things will hang out with each other, and that "birds of a feather flock together"....sometimes. What "feathers" we care about vary from person to person. And the results can be tragic if we're not careful. Human nature biologically needs to categorize things, to make some sense of the world around us, to find patterns. Too often, we will find patterns that aren't really there and that gets us into trouble.

It's often personality-dependent, as in, some of us will care about skin color from birth, some of us won't. Some of us will hate each other because of the neighborhood we grew up in, or our favorite sports team, or political affiliation, or boldness of clothing, or jewelry, or makeup. We are naturally leery of anything and anyone that's "different" from us. And that's very often caused by all of our biology and other factors we aren't aware of. I may not judge you by your skin color, but I'll be more comfortable with you if you talk like me. As we grow up and teach our children to grow up, we can make the world a better place by teaching them the advantages and limits of our diversity. A unified group can accomplish a great many wonderful things working together, but a diverse group is much more resilient against the changes the environment throws at us.

Incidentally, my wife says I'm most comfortable around the stereotypical southern California surfer. And she's right. When we were dating, I took her to a beach near my home and we sat on the sand on a blanket talking about life and whatever else we felt like. The topic turned towards the "California Stereotype." You know, the one that says all surfers say things like "Dude" and "Gnarly". I had just finished telling her that that stereotype is totally overplayed and overdone and hardly anyone actually talks like that.

As if the universe were trying to prove me wrong, within 2 minutes of that conversation ending, a pickup truck pulls up behind us, two guys who look like they'd just gotten off work for the day get out and start getting ready to head out into the water, putting on their wetsuits and getting surfboards ready. Listening to them talk sent my wife into hysterical laughter and I had to admit she got me. We were hearing things like "Dude! Look how sweet the sets are today!" "Oh man, you shoulda seen it yesterday! There were all these gnarly barrels and tubes! I totally got wiped out by this monster slab! It was frikkin' awesome!"

She won that one. Bronwyn:1, Christian:0

Later, we stumbled across a news interview of a surfer describing how awesome the conditions were at a beach in southern California on a particular day. The surfer used only a few words the average person would understand and has become known as the "So pitted" guy. He's a hero of mine now.

Bronwyn 2, Christian: 0.

Stereotypes exist for normal reasons! Daniel Kahneman and his colleagues performed a set of well-known experiments that showed this. The "Tom W." and "Linda" experiments showed that the large majority of people, even graduate students at prestigious universities, who understand statistics, probabilities, and psychology will still stereotype and "profile" others based on superficial qualities, against logic. Even with the expressed warning to the test subjects that the description of Tom W was unreliable, most people lumped the self-centered, pun-enjoying, nerd as a Computer Scientist. Or that even though it would actually be LESS likely, the Social Justice Warrior Linda was assumed to be a feminist bank-teller as opposed to a "regular" bank-teller among the choices people were given.

There are patterns across cultures and groups of people that hold true at least PART of the time. Our brain will naturally make those shortcuts throughout life, and fighting against our own prejudices will be a life-long battle we have to wage with ourselves while staying a part of a social and interconnected world. We will always feel threatened by "other" groups of "those" people. That being said, even if another group of people is threatening us, we don't have to FEEL threatened by them. As a personal practice, I tend to mistrust anyone or anything that tries to lump bad behaviors into a group of "them" or other people. The popular phrase for it now is "Tribalism." The lumping of people in compartments and well-defined groups as if everyone in that group acts the same way or believes the same things. Yes, groups of people can be responsible for destructive actions or of wrong ideas, but blaming the Group for an individual's wrong ideas never helps the situation, though it can help us understand where it came from.

Our brains will work better when we are NOT afraid of what is threatening us. I always loved the porch scene from M. Night Shyamalan's "The Village". When asked why he is not afraid while the rest of the village shakes in their boots, Lucius replies, "I do not worry about what will happen, only what needs to be done." When we are scared, we become drunk on adrenaline. The calm voice of a prefrontal-cortex gets drowned out by the brash, less-thoughtful shouting of the inner reptilian brain. And we don't have to be whipped into a mob-frenzy with looting and rioting on our minds to get drunk on adrenaline. The anger and passion that gets stirred up inside of us while we jeer along with a political commentator —who skews reality for laughs and ratings— can stick with us for months and years. The injustice that a talk radio show host tells us about can leave a long-lasting imprint on us that can take conscious effort and work to undo later.

The law of nature is "Kill to eat, or kill to keep from being eaten." That still carries over into our lives, even though very few of us are killing or being killed. We are often scared and afraid of anything that's new to us. In Rudyard Kipling's, The Jungle Book, Shere Khan, the Tiger that the rest of the Jungle was afraid of, was afraid of Mowgli, the little "man-cub" boy. Mowgli posed no threat to Shere Khan, but Shere Khan was afraid of who Mowgli would become. Yes, the ferocious killer, was the one that was afraid.

We become afraid, not of what is happening right now, but of what might happen. Dale Carnegie wrote an entire book on this, "How to Stop Worrying And Start Living." Worry is fear. We become afraid of others who are encroaching on our way of life even if they are not threatening our lives. That is why we are afraid of other political parties, of other races, of other countries and nationalities, of anyone who looks different from us. If there is anything unique, different, or distinguishable about our appearance, someone will judge us on it.

Daniel Kahneman calls it, "What you see is all there is." It's the shortcut our brain makes. It has to start categorizing everything it encounters, and when all it knows about something is the appearance, it will instantly start making assumptions based on it. Especially when we are scared, nervous, awkward, apprehensive, worried, or uncertain. These are times when our brain feels the pressure to figure everything out in a hurry. Know that anyone who judges you on your appearance is scared, to some degree, in some way. We can take advantage of that for our own selfishness, or we can accomodate and compensate for that, as a favor to the other person, by altering our behavior ever so slightly, to help them feel safer.

When we feel safe, we will put up with a lot. But feeling safe is a joint activity. We need to work to make ourselves feel safe, and we need to work to help those around us feel safe. We must find ways to make life safe for other people and ourselves. I love the little incident R.J. Palacio describes between two characters in her fictional-yet-realistic book, Wonder, where the boy with the malformed face, August, is eating lunch with his new best friend, Jack. August has been extremely self-conscious of his appearance and he is very aware of everyone else's fear of him because of it. He had never attended a regular school before this year because of his fear of everyone else reacting to his appearance. So when Jack says to August he should get plastic surgery to fix his face, rather than lashing out at Jack, or withdrawing away from him, he giggles and laughs it off and says, "Hello? This IS after plastic surgery." August felt safe with Jack and wasn't upset by Jack's bold statement.

There are numerous studies that show that when patients who are recovering from surgery are given control of their own morphine, they will often use LESS morphine to dull their pain than the doctor or nurse would give them. When we feel in control of our surroundings, we are less afraid. When we don't know when our next shot of pain relief is coming, we become nervous and anxious and worried that it won't come in time, and we'll be left to suffer unnecessarily. So, when we are in charge of our own pain relief, we KNOW when it's coming and that we can do something about it at any time. And we know that morphine isn't good for us, so we'll use as little as we really need, and in the end, we will use less.

I've seen this in dentistry. Nobody likes the process of getting dental work done. I get it. I get told it every single day, "No offense, but I HATE the dentist." I hear ya.

Everyone comes to the dentist a little bit nervous, or sometimes, a LOT nervous. So nervous they can't think straight, or rationally, or even think at all.

I tell my team: our job is to help patients feel safe in our care. We let patients be in control. We offer choices, we never force anything. We always give them an 'out'." I let my patients know, "If I'm doing anything that's uncomfortable or bothers you, just raise your hand or make a noise, we're good at taking hints." And the more we tell this to patients, the fewer of them actually ever need us to stop. They are in control and they won't stop us until they have to.

My hygienists do this as well when using the ultrasonic scalers to clean teeth. It sprays lots of water and can quickly fill someone's mouth up, uncomfortably full, where they'll likely dribble water all over themselves. When I clean their teeth —or when my hygienists do, I'll teach them to say—, "This sprays a lot of water. I'll give you the 'straw' so you can slurp the water up. I'll give you a break every couple of teeth, or, if you feel you need it at any time, you can turn it on and I'll take it as a hint and get out of your way."

This makes them feel in control, and thus, safe. Then they put up with a lot more than they normally would. Their fear is gone, they are thinking clearer, and they can get their health needs met quicker.

Addictions, bad habits, violent behavior, mean-spiritedness all fade when we feel safe. Like Dale Carnegie would say, "No one kicks a dead dog." When we are not afraid, we don't lash out or try to escape.

When we know when we will get sex next we will wait longer for it. When we know that food will come we will put up with hunger. When it doesn't show up as promised, we get scared and anxious. When we have our own safe space, our own territory, we feel safe. When that territory is encroached on, we become afraid. When we feel safe, we are more likely to explore and try new things. When we are afraid we stay holed-up in our own self-made prisons.

Men that don't feel threatened by a female coworker, or superior, will behave very differently from a man who feels threatened by a woman in the workplace. Women (and men) who feel safe at work and that their efforts are validated and appropriately rewarded will almost always work even harder on the merits of their own drive.

Preaching and shouting at each other to be nicer, just doesn't work. It stirs up fear in those we are shouting at, and they will retaliate. We just have to go out and actually BE nice. We need to do things that will help others feel safe. Motorcycle bikers were getting a bad rap because of groups like "Hell's Angels." What did other, innocent, bikers do to reverse the prejudices they were facing? How did they change the profiling that was going on against them? Did they riot in the streets about how ill-informed everyone's opinions of them are? NO! They formed service groups, like Bikers Against Child Abuse, and people happily started spreading their stories across social media.

We cannot force people to change their minds because we say so. That scares them. We can only invite them to enjoy our company by doing what is right and hoping they see that. Our attempts to change others and make the world a better place by shouting at others arises from our own fears about them. We all feel justified for our own opinions and prejudices while condemning others for theirs, which they feel justified in feeling. Are the far-right and the far-left ends of politics justified in being afraid of each other? Probably. A woman who was mistreated by a father or husband, or multiple partners, will often form a fear and dislike for men. That's totally a normal response. It still doesn't mean all men are like that.

We have to teach people to recognize fear inside themselves, only then will we make headway against racism, sexism, violence, dislike, animosity, and hatred.

If I'd grown up as a white child in other parts of the world such as central Africa, or Asia, or the Pacific Islands there's a good chance I would feel racism and picked on and despised by those around me. If I tour around Europe with a big American flag on my backpack, I would be treated differently than if I were showing off a Canadian flag. Many parts of the world perceive Canadians as more peaceful than Americans and make blanket, profiling assumptions about people on a backpack patch alone.

Even living in Canada as an American volunteer, we would get targeted, followed, yelled at, insulted, and objects thrown at us from cars for no other reason than our uniform. The strangers we would interact with would very often insult us and threaten us simply because of our nationality. The same goes with black people in large parts of the US, or Asians in certain parts of the world or Native Americans in certain parts of the world. We will all feel the effects of racism, sexism, or various other -isms, for good or bad, if we ever live anywhere that we don't fit in.

I've heard more than one person tell me that, because Americans are perceived as a violent culture, when violent crimes occur in Asian countries, the white Americans are profiled and suspected first. In Japan, it's not uncommon for a plane from the U.S. to deboard and the non-Asians searched more thoroughly by security, while none of the ethnically Asian passengers will get searched.

Because of our sex-filled media that pours out from the US to the rest of the world, American girls are seen as "easy" and promiscuous by people in many other countries and targeted by predatory men, for no other reason than being an American woman.

Jared Diamond, in his book, "Guns, Germs, and Steel" describes the animosity and racism that has existed for the entire history of humanity, across the world. He even gives a personal experience of the strong racism among four different people, from four different ethnic groups, living in the same region of Papua New Guinea, all native to the island, who were helping him with his research, and how those ethnic groups came to be in the same place, having to interact with and deal with each other. His amazing research goes incredibly far to show that success and privilege has less to do with race and more to do with WHERE we live, or where our ancestors came from.

There's nothing very special about Caucasian Genes, but there were certain factors built into the environment of Europe that allowed Caucasians to thrive as well as they did. There were also factors built into the African, Asian, and Australian environments that allowed the ancestral people of those regions to thrive or be held back, purely out of what plants, animals, climate, and features dotted the landscape. There is no superior race but racism is still prevalent. Privilege is a real thing and very dependent nothing else then where we were born and who raised us. However, if we blame them for all of our problems, we are giving in to instinctive nature instead of using bravery and courage to rise above it.

If our bad experiences are repeated over and over and over we will form feelings (whether we want to or not) against those who are persecuting us. We only have to touch a hot stove once to learn to stay away. We only have to get hurt a few times by others to learn to survive by staying clear of them. Our avoidance of groups of people, our tendency to profile, comes from repeated experience. So, we need to be extremely careful that we allow for exceptions, that we realize the limits to our experience.

We need to understand that our perspective on the world isn't complete, and in fact, may be skewed and distorted big-time by our profession, or lifestyle, or family, or simply by where we were born. Just because you have had a repeated type of experience with a group of people, doesn't mean that everyone belonging to that group will be like that, or that your experience is unique to even that group. Every group of people on this planet can be mean, and rotten, and awful, and they can all be pleasant, caring, and gentle, depending on the circumstances.

I'm not justifying stereotypes, profiling, racism, hatred or animosity. Not at all! I am saying that if we don't recognize what triggers those natural instincts, then we can't actually solve it. I don't believe we'll ever get rid of any of them completely, we will just have to do our best to teach each other to know what fear is and to watch out for its effects. And we will have to teach every single new generation that's born to this earth.

In the song "I'll Follow You Into the Dark," by Death Cab for Cutie, the singer describes leaving his school because his teacher taught that Fear was the "Heart of love." While I can see why some people would think that or feel this common idea, the real truth is that Fear is the opposite of Love. We have to learn that things have their opposites. If we want to survive in this world, we need to learn the difference between Pleasure and Pain, Health and Sickness, Light and Dark, Fear and Love. Even many religious people have no problem personifying "love," or "light," or "goodness" as God but seem allergic to personifying Fear and Hate as the Devil. Maybe some of us are afraid to give our fears a name, for fear of empowering it. Like, "He Who Must Not Be Named" a.k.a. Lord Voldemort in J. K.Rowling's Harry Potter books. Such avoidance only perpetuates our fears and makes them stronger.

We need to recognize and confront fear. We have to. We have to chase it away with love, and knowledge, and light. It's never easy, but we have to recognize the source of our fears and call them out. Bring them out of the corner we are avoiding, and shine the light on them. As we practice this, just like when we practice anything, we will get better and better at it.

When we are afraid of pain and sadness, we take drugs. When we are afraid we won't ever receive enough love or sex we go looking for it in destructive places. When we are afraid of losing our partner or children, we become co-dependent, trying to manipulate and control them. When we are afraid of being hurt, we preemptively hurt others, all the while feeling justified

Most people are okay with the idea of self-defense. Of hurting or killing others to keep from being hurt or killed ourselves. Self-defense isn't so cut and dry, though. If we KNOW a threat is coming, are we justified in preemptively striking first? Do we turn the other cheek seventy times seven? What if we make a mistake and take the first strike only to find out the threat wasn't as great as we thought? What if we let someone hurt us time after time expecting them to eventually stop, and it becomes an abusive situation, that will never stop?

How threatening something is, depends on our own level of fear or control in a situation or perceived intent. Someone that threatens to hurt us and has the power to do so, should be treated as a threat. Someone who keeps hurting us despite promises to stop should be treated as a threat. We should try to remove threats from our life, either by fighting, withdrawing, or through careful and clever diplomacy, tact, and conversational skill. Sometimes, we just need to live with a scary threat that isn't doing anything to us.

Sometimes, we can't get rid of something that is harming us, so we learn to defend ourselves from it and protect ourselves. Very often, with some amazing skills, we can turn those who want to hurt us, into people who want to protect us and help us. Not always, but often enough that it's worth trying. There are incredible diplomats, senators, presidents, CEO's, and Chairmen and Chairwomen of Non-profit groups that have amazing interpersonal skills and manage to build relationships where most people would fail. This really is the most valuable skill anyone can learn.

Almost all of us will accept —as a necessity— the idea of killing of an enemy soldier, but only a few of us nowadays can accept the idea of killing, or sentencing to death, a repeated murderer. It's a tough decision to make. And I will not use this book to promote one way or the other. I'm at peace with how most of our laws are at this moment, but I wanted to give an example of how fear changes our standards. We are afraid of an enemy soldier, we are not afraid of a psychopath locked up in a maximum security facility.

Now... fear can be a great motivator, to do good. Fear of punishment can help us avoid criminal activity, though as history shows, it's not a very effective, or long-lasting motivator. If we want to break a law or do something wrong, we will eventually find some way to justify it in our own minds, and our fear of punishment or getting caught will be ignored. At times, though, fear of doing wrong, can keep us doing good,

We must be careful with this justification. Living your life by avoiding what you're afraid of is like driving your car around in Reverse. Just because I can drive my car the right direction while it's in reverse, doesn't mean it's a good idea, or that I should do it often. Driving a car like that is harder to control and is more likely to lead to an accident or destruction of other's property, even if you started in the right direction or had good intentions. At some point, it's best to turn the car around and drive forward the way it was designed to. So it is with fear. In temporary moments it can improve our wisdom, or keep us out of danger, but we can't live our life based on fear. We will begin to avoid things that are good for us, for fear of pain, or fear of failure. That fear will cripple us.

Even when we've labeled our fears, we then have to go through the struggle of fighting against it. That is a necessary part of growth. Love turns fear into something useful and wonderful.

Let me get back to the idea of fear being necessary for love. Some feel that because fear is the opposite of love, love is dependent on it to exist. We may not know or understand one without the other, but they are independent of each other. For example, without black, it's harder to understand "white". Without light, it's harder to know darkness. Without fear we often don't appreciate Love. Light is substance, darkness is nothingness. Darkness will always exist independent of any light source, and light exists on its own whether or not it is surrounded by darkness. Love exists all on its own and is the most powerful tool to combat fear, and where there are love and light, the odds will always play in their favor.

10

Introduction to Part II - An Introduction to Love

The rest of this book will cover all the strategies and tactics and steps to overcoming fear. There are many forms, programs, and systems, and they all go by different names. Christianity calls it Repentance, Salvation, Conversion, or being "Born Again". Eastern Religions have Enlightenment, Nirvana, or Satori. Rather than reinvent the wheel, convert you to a religion you don't belong to, or try to explain cultural backgrounds to get the point across, I am going to use the most basic and simple versions of these ideas that just about anyone can grasp. The process of overcoming fear, and addiction, and achieving long-lasting inner peace, progress, and knowledge is very well captured in the 12-Step programs. These programs are not meant to be a one-time cycle, they are the principles for achieving self-mastery and the ability to free our mind to learn and grow and heal so that we can work towards whatever we want to work for. Use them over and over and over again as often as you can or as is needed to achieve health and long-term happiness.

Following an actual 12-Step program isn't the only way to succeed, and just because a program is framed around the 12 steps, doesn't mean it will work for you. The next large section of this book will talk about various other programs and methods that people have used to achieve success at self-mastery. It all depends on how well these principles are applied by those teaching it and those using it. I know too many who were disappointed by the condemnations or judgmental members of a group. Maybe it was the hypocrisy of the Facilitator or other attendees of an AA meeting. Maybe they didn't fit in with the culture of their religious organization, or church, or community group, or whatever. That is going to be a fact of life.

We are all just as hypocritical, judgmental, and condemning of others in our own unique way as those we point the finger to. The faster we all accept that and move on, the quicker we can heal. Self-righteousness breeds self-righteousness and is based on...fear. Fear of judgment, fear of failure, and fear of stagnation. Don't be angry at others for their imperfection, just accept that everyone is imperfect and that you need to work on your own while they work on theirs.

Whether we make it where we want to go will depend on what life gives us, what our backgrounds are, what our physical, emotional, and mental limitations are and what we choose. The end-goal isn't necessarily perfection, but if we aim for perfection, we can only get better, and we'll bring others along the way.

Carl Shurz, an American ambassador to Spain in the 1800's, said it well, "Ideals are like stars. You will not succeed in touching them with your hands, but like the seafaring man on the desert of waters, you choose them as your guides, and following them, you reach your destiny."

Wayne Gretsky is often quoted as saying, "You miss 100% of the shots you don't take."

Keep reading, and keep trying.

II

How to Make Change Happen. How to Love People, Including Yourself

The remaining chapters will focus less on science-y kinds of things, and more on what to do about it. Strategies for controlling your body, mind, and spirit. You can read this section alone, or in order. It will help you either way.

11

You're Just Not Good Enough

I've got a friend, Steve (not his real name), who has had a rough time. He and his young wife were having a lot of troubles. They were both very career-oriented. He worked towards a prestigious professional degree and she was very entrepreneurial. He worked full-time and took classes on the side. She kept bouncing from one career or certification or business lead to the next. Seafood importing, fitness instruction, retail. They were from different countries and had very different ideas of what they wanted out of life. She wanted fancy cars and outward prestige. He wanted a down-to-earth lifestyle and to raise children. For several years they did their best to work with their differences. There was love between them.

They did have many things in common and I watched them both try to amplify and focus on their similarities and shared likes, and I know he worked hard to respect what made her unique. I imagine she tried to do the same to the best of her ability. A major source of conflict and hurt between them was that she no longer wanted to have children like she did when they were first married while he couldn't wait to be a dad. We became friends when they moved into our neighborhood and realized we had a lot in common, like our love for Subarus. His WRX was much cooler and faster than my Legacy Wagon, and we would talk while driving his up and down the mountain roads and canyons near us.

His wife would leave for months at a time pursuing business ventures. Alone a lot of the time, he did his best to fill his spare time in productive ways but would often turn to pornography and masturbation, which he knew he had become addicted to and he had a hard time avoiding. It began to impact his daily life and the demands and responsibilities he needed to fulfill for work and school, so he reached out to others for help. He began attending a 12-step support group. He got counseling. He read many, many books. He confided his struggle to his wife. Scared, insecure, and hurt, she would lash out at him in angry outbursts. They would argue and fight. Their arguments would often turn into arguments about their differences in money or their different goals in life. They often become very heated. They both tried to keep level heads, but often failed.

Eventually, they divorced. He cleaned up his private life and began to move on with life and moved away, though we stayed in touch as best we could. He did his best to make up for his faults. He figured out why he had behaved the way he did and made serious changes to avoid them in the future. He kept pursuing his education while working to pay for it without debt. With a sense of confidence that came from knowing he had done everything he knew how to do, and had fixed the problems he previously hadn't known he was causing, he began dating again.

He met a beautiful and smart woman who wanted the same goals in life that he wanted, among them a family. She had a career in the same field he was interested in. She had a similar work ethic he had. They got married, and at the last I heard, she was expecting a baby and they were both incredibly excited about it. On the outside, they seem to have a much happier marriage than his first one. Given what I know of his growth and changes, I imagine it's not just superficial and there is true happiness in their marriage, even though it's not perfect by any means.

When you've accidentally ordered... or when life has handed you a big batch of lemons (let's all be honest, life doesn't give you lemons, it gives you lemon-flavored dog-poo) you can't always do much to make lemonade out of it, but it will make for good fertilizer to grow a new you. Anyone who's dealt with an addiction, a bad habit, a compulsion, an obsession, a mania, a personality flaw, or any other imperfection of any kind that has had a negative impact on yourself, or your loved ones and neighbors and community, will learn that the damage has been done. You can't go back and fix it, and you can't always make it better. Sometimes you can.

Either way, you CAN stop it where it is and make a new life for yourself, slowly, starting now. Fortunately, only a very few people have to start life over from scratch when dealing with or confronting life's problems, though those who've had to and made it through, will tell you, often it was the best thing that ever happened to them. It's not always about turning lemons into lemonade, it's often about learning to enjoy lemons. There's the phrase, "Life's not about waiting for the storms to pass, it's about learning to dance in the rain." Something like that.

Unless you are omniscient, omnipotent, and omnipresent, chances are you need some help. You are incomplete. You are lacking, deficient, flawed, weak, etc, etc. My thesaurus has more words I can use, but I think you and your beat up self-esteem are getting the idea.

The idea that we all need help is just one basis for just about all of the world's religions. The behaviors that keep people from God, or perfection, or completeness, harmony, inner peace, etc are called, "Sins" by many people. Basically, a sin is any imperfection or flaw in ourselves, whether due to action, or in-action. There are the obvious "sins" like the breaking of the Ten Commandments, and various other admonitions written down in the Bible, Tanak, or Quran, but of course, there are a billion other ways to mess up and "sin." Christianity teaches that we are ALL sinners and all need repentance.

So, thinking of it in this way, we all need recovery from something. Whether it's recovery from our smartphones, our Ben and Jerry's Ice Cream binges, the latest season of a Netflix show,...whatever, we all need help overcoming all the biology I just spent the first half of this book explaining. Hindus also believe that we all need improvement and that by working to improve ourselves through reincarnation after reincarnation we can one day achieve Nirvana.

Buddhists also assume that we are all imperfect and they will spend their lives working to achieve Enlightenment and Inner Peace. Of course, each person's religion only works as well as their level of life experience, and the thought and effort and self-introspection they put into it. Too many members of every faith (even atheism and agnosticism) can't see their own imperfections because they are too busy looking at everyone else's, blissfully unaware of the damage that they are causing.

We all think to ourselves, "It's not that bad. I'm not as awful as 'those' people." Except that, yes, we are. Yup, even you Mr. or Mrs. Self-Righteous. Can you be self-righteous if you're not spiritual or religious? Absolutely. And chances are, you already are self-righteous. We all are. We're all hypocrites at some level.

There's no shortage of people who know that this world we live in is pretty messed up, and it's messed up because of us. I think we all agree that we have a lot of room for improvement as a species. Where many of us fall short is realizing that we all have room for improvement as individuals. Change is what OTHER people need to do, right? It's what Humanity needs. Me? I'm okay, I'm a good person. My flaws aren't that bad, but those OTHER people? They are crazy!

I remember having dinner with a good friend, an excellent chef and cook. We're sitting in his kitchen while he gets dinner ready to have a bunch of friends over. He's slicing up some pork chops and the topic of hunting comes up. He's sitting there, pork chop in hand, slicing it into finer pieces, more pork chops on the counter, and he says, serious as anything, "You know, I just can't stand that. Hunting. I mean, how do you go and kill another living creature? I just don't agree with that at all."

Now I don't want to analyze everything that went into or out of that statement. I didn't say a thing, I just held in my surprise, and let it go, and stayed and enjoyed some delicious pork chops with some really good company. My point is when we are removed from a situation, or task, or environment, it's REALLY, REALLY, REALLY easy to make condemnations and judgments on other people. Super easy. So...let's hold off on our judgments and just take the assumption that most of us are pretty decent —most of the time— and many of us would do the same or similar things if we were the same person, from the same background, and the same upbringing, in the same circumstances.

So: Given that we all have a lot of improving to do. How do we do it? Where do we start? And will we ever get better??

I'm glad you asked. Read on!

12

When You Lose All Hope, Breathe

Sometimes, something happens and you will find yourself having lost all hope. You will feel like giving up at times. You will feel so awful and miserable and overwhelmed. You will feel like things will never get better. You will want to run away. You will want to quit.

Your heart will race like a car with no brakes, careening down the hill. Your mind will scatter like sheet lightning across the sky in a never ending rumble of thunder and flashing anger, fear, and panic. Your breathing will often get real short and fast and you'll feel like you just can't get enough air. Like you are going to drown. You'll want to cry your eyes out in despair or scream in rage at the sky and not know whether to do one or the other so you'll go back and forth between the two, quickly.

You will feel like no one will miss you. Like you are a burden to the world. Like you are worth more dead than alive.

You will feel like it's all everyone else's fault, or like it is all SOMEONE else's fault. Maybe you'll think it is God's fault, or your spouse's, or your boss's. Or your parents, friends, or neighbors. Maybe it will be your church leaders, or teachers. Maybe it will all be the fault of the stranger or trusted confidant who abused you or misused you when you came to them for help.

Or the twisted person who took from you what wasn't theirs.

Or, you'll settle on the idea that it is all your own fault and that will make you feel worst of all.

When you feel all of these things and more, pay attention. When you feel like you are losing control and it is all hopeless that's when you need to perk your ears up and realize what is going on. Step outside yourself for just a split second and look at what is happening to your mind and your body in those moments.

Realize that those reactions you are having are not YOU. They are the instincts, and physical reflexes, and the screwed up biological drives gone haywire. An inner-reptilian-brain gone nuts. You have been dealt some kind of trauma and your outer brain no longer has control. The inner brain has lit up like a fire-cracker and it won't stop until it runs out of gunpowder.

Find a safe place and let it run its course. Let it burn out. Don't make any lasting decisions while it is going. Don't let yourself or anyone you love get caught in the crossfire. Run home to your spouse, or if you don't have one, to your parents, if you need to. Find a quiet room and just let the steam blow itself out. In between volleys of bullets of out-of-control emotions, run for shelter until you can take control again.

And how do you take control of your body when you've lost it?

Breathe. Just breathe. If you can control your breathing, you can control your world. If you can't control your breathing, focus and concentrate and keep trying until your breathing is your own again.

This breath of life. Air. Oxygen. Without it, you'll lose complete control over your body in about 5-6 minutes. When it starts racing, when you're in a panic, slow it down. You may have lost control over your instinctive primitive brain, like a wild monkey. You may have lost control of your heart which is thumping away like a time-bomb. But your breathing?

You've still got control over those muscles that lift and raise your rib cage. You can still take back control over your diaphragm, that sheet of muscle between your lungs and stomach.

If you can control your breathing, you can learn to control anything else. Slow your lungs down, and you'll slow your heart down. Slow your heart and you'll be able to slow your mind. Then, once you have control over your mind, you can control the rest of your body, and from there on you can learn to gain control over the world around you, a little bit at a time.

It all starts with learning to breathe and control your breathing.

So many of the martial arts concentrate on breathing. Yoga is centralized on breathing. Much of meditation is focused on learning to control your breathing and simply manage it for as long as possible. Learning to count to 10 before you lose your temper is an attempt at getting you to take a breathe before you act. Free-divers, who manage to dive to hundreds of feet deep on a single breath of air for 5 minutes at a time, have to learn to overcome those panic-inducing urges that make their chest contort in an attempt to suck in air. Instead, they use those urges like a timer to keep track of how long they've been down.

My kids have learned that when they are feeling car-sick, they need to open the windows, get the vents blowing fresh-air on their faces and think about their breathing.

Anyone who's ever tried to camp or sleep in a small enclosed space will know the headaches you get when you're breathing too much used air and the Carbon Dioxide starts to build up.

Sleep Apnea is a medical condition that often occurs when people hold their breath in their sleep. It's usually caused by a person's tongue and throat relaxing so much that it blocks their airway. They can weigh enough that it can actually overpower the normal, automatic breathing reflex. Only when their heart begins racing, the carbon dioxide has built up, the oxygen is running low, that the brain stirs the person awake, just enough, to take a breathe. Spouses and partners will often hear the patient stop breathing for a minute at a time and then suddenly take a big gasp of air. Snoring and grinding of teeth go hand-in-hand with it quite a bit, as well.

This condition grows slowly over time, is very heavily associated with sleeping on your back and being overweight, and causes a WHOLE number of problems. First is that you just aren't sleeping well. No one remembers waking up to take that breath, but it jerks them out of their deep R.E.M sleep that's so necessary for a healthy mind and body. I've seen reports showing heart-rates race up to 150 to 180 beats per minute. That's like a super intense workout! While you should be sleeping! Over the years, that stress and lack of rest leads to all sorts of health problems, like diabetes, heart-disease, strokes, auto-immune issues, mental health problems, etc, etc. It's a slow killer. Breathing is important!

I had a scout-leader as a teenager who was a chiropractor and also a master of Judo and Jujitsu. He was a fairly short, middle-aged Japanese guy. I remember him giving us a lesson in self-defense. Among the tips he gave us, were to slow down your breathing when you are in a match. Then we watched as he took on 4 high school football players and wrestlers, back to back, without a pause before he even started breaking a sweat, while they all crawled away from the mat totally defeated.

Another time, as we hiked Half-Dome in Yosemite National Park, he caught up to me, saw me struggling just to keep one foot in front of the other as we neared the summit and its terrifying cables. "Focus on your breathing. If you need to, breath out for as long as possible, then suck in as much air as you can as fast as you can. Do that for as long as you can think of it." Then he marched on by, practically whistling a cheery song as if it were a stroll in the park.

If you ever fall through the ice into the cold water below, every survival expert will tell you the first thing to do is slow your breathing. Don't let panic take you.

I'm tempted to go into all the science of how this works and why it's so incredibly AMAZING! But, my wife thinks I should spare you the details, and she's probably right. I will say this, your body is terrible at measuring how much oxygen is in your system at any time. There's not really a warning system for low oxygen, you just black out before you know what's happened.

Remember the story I told about Destin Sandlin and the NASA chamber? Because he was able to "breathe" and let go of the CO2 in his system, his body didn't give him ANY warning signs that anything was wrong.

It's REALLY good at measuring how much Carbon Dioxide (CO2) and used air is building up in your blood. You can actually survive quite a bit of CO2, but your body starts screaming at you at the slightest buildup. You have about 3-5 minutes of oxygen in your body at most times, but it only takes a single minute's worth of CO2 to make most people panic. (Still, don't practice this underwater without a buddy system!!)

What's the lesson? Your body will very easily panic when it doesn't need to. Your outer brain, the part that just learned what it did about oxygen, can be MUCH stronger than your primitive brain, if you practice controlling it. You can survive much more than you think you can. I promise. Almost nothing is as bad as your panic will make you think it is. Your body is tuned to look for problems, but not things that are going well. That takes practice. Gratitude is not a natural instinct for most people.

So, when life gets out of hand. When you've been dealt a blow you just can't handle. When you feel like you will never recover. When you feel like nothing makes sense. When you feel like you are going to be miserable forever, don't believe it! Just don't! You've got more "air" in you than you think.

When you've lost all hope for the future, rely on the habits you formed when your mind was clear. Keep the promises and agreements you made to others even if you don't feel like it. At least as many of them as you can muster, or even just the few essential ones. Let auto-pilot keep you going while you need it. Go ahead and be a zombie or emotionless robot for a while. It is okay.

You've got a lot going on and you don't know how to process it all. Step back and focus on the basics. Let your past momentum keep you moving forward. Most times it will carry you right on through the storm. Like a mountain biker splashing through a rocky river or stream, sometimes you just need to hang on tight and just plow through. If you crash, you crash. Hope that you don't, but deal with that if it happens. Let momentum carry you as far as it can.

When you catch yourself losing control, breathe. It is the one thing you can ALWAYS fight to take back control of.

And stay as consciously in charge of your breathing for as long as you can. Your body will fight to let the automatic systems take control over it, but don't let them! Stay as focused on the air going in and out of your lungs for as long as you can and slow it down. Then, when you've got your breathing under control, work on the other things.

So, like Anna Nalick sings, "Just Breathe."

13

Desire, Hope, and Hacking Biology

The first step to change is just wanting to change. Wanting to BE better, wanting to GET better. THINKING that something can change. If you don't have this desire or this hope, nothing will happen. The phrase, "faith as a mustard seed," is very appropriate. Just like a seed falling into fertile soil grows into a large bush or tree, a simple desire, want, or hope is all that is needed to start lasting change.

What if you don't have a desire to change? Lots of things can give us a desire to want to change. Personal motivation, social motivation, or the environment around us can motivate us to change. Yet, we get to choose what will motivate us or not. Sometimes we just need to get out of the way and let the other things in our life spark the change in us.

What if you want to change but don't have any hope that it's possible? I hope that reading the first half of the book showed you that change is always possible, often it's just a matter of figuring out our biology and "hacking" it. If you're still not convinced, there are other ways to receive that hope. Find someone who has gone through the change you want to apply to yourself and set up a time with them when they can be completely honest and let them tell you their story of how they changed. Find a support group filled with others who've gone through that change, or led by a facilitator who has undergone that change. How about a church or religious setting with others who've felt the change in their life for the better?

But why would we want to change? Why go through all the trouble? It's tough work. So tough that so far, only humans have been able to accomplish it, and even just barely, on individual levels. Why change our life, our lifestyle, or our environment, just to become someone else? That will be among the most difficult decisions of your life you will ever make. But here's a carrot to dangle in front of you.

Summary: Be decisive.

14

Happiness, Gratitude, and Whitewater Rivers

What makes us happy? What is happiness made of? Where does it come from?

There is a large collection of TED talks you can find online, full of scientists who can tell you what leads to happiness and how to get more of it into your life. And they'll agree with many of the pastors, priests, monks, or rabbis you've come across, too.

First off: Life makes no guarantees. The world doesn't actually owe us anything. Just because we are raised in a certain place, surrounded by others who get to achieve certain life goals does not mean we will get to achieve the same things. Long life is not guaranteed. Health is not guaranteed. Wealth, family, relationships are not guaranteed. The American dream, raising your social status, owning a home, retirement, are not guaranteed.

We get these things either through hard work, or as gifts from others. Even then, bad timing, immaturity, poor decisions, the actions of others, or nature can rob us of anything. Crying "Foul" and giving up is the natural reaction. We think it's unfair, and maybe it is. The problem with the world is not that things are unfair, it's that the world is much more fair than any of us want to admit. Stuff happens. To all of us, at different times, and in different ways.

Comparing ourselves to others and their privileges, blessings, or comforts is cancerous and destructive. It's nice to have goals to aim for and we should have goals to aim for. But we are in a race only with ourselves and should only compare ourselves to our self. This is much, MUCH easier said than done, and it is crucial for happiness and fulfillment.

Much of our depression, anxiety, and worry comes from thinking that our life SHOULD be better than it is. We feel sadness and anger and we think that we shouldn't feel those things. We have trouble sleeping and we worry that something is wrong with us. We experience loneliness and fear that there's something deeply wrong with us. We struggle with poverty and worry that we will have to scrape by our whole life. We become afraid that it will never get better. We lose Hope. We give up trying. We can't figure out how to fix what is wrong with us so we quit, or stop caring.

Happiness begins with a decision. First, you DECIDE to become happier. Why you'll make that decision is hard to pin down, it'll be different for each person. Once you've decided to have more happiness, you'll next need to know what to DO. Happiness requires action. It does seem to come "naturally" to some people and is more difficult for others. What we find when we look at their lives closely is that usually those people who are "naturally" happy have been taught good happiness-inducing-habits since they were born, and were raised in a home with good happiness-inducing habits and lifestyles. Others have to work hard to become happy and develop those habits. The greatest "privilege" someone can be born with is this trait. Some are born into it and others not.

A safe and stable home where happiness —and the behaviors that lead to it— is taught is a tremendous privilege to be born into. Poverty is nothing to celebrate, but extreme wealth does little to make us happy, either. Many

Lottery winners have been followed up on and it's discovered that very few winners of large lotteries would rate themselves as any happier after their windfall, than before.

Several studies have been done on what the happiest income is, and with the varying amounts that the studies come up with, the happiest incomes are usually just a little bit above the average for the area. In other words, we are happiest when we we have just a bit more than our neighbors and peers. When our bills are paid and we have a little bit extra. The natural tendency is to add more bills to our life and raise our lifestyle to a new level. Then we have to compare ourselves to a new set of peers that we want to be like and be just a bit better than.

Very few of us can resist the temptation to out-earn our ability to spend. We are perpetually stuck in the need and desire for more. It's in our nature, it's a drive inherent in many of us. That drive can sometimes get shut down when we fail to move forward after years and years of frustratingly slow progress. When we feel stuck we eventually give up and settle. The drive for more and for ambition is a good thing, to a degree.

We can maintain our ambition while being grateful. It is those two things combined that help us find the greatest happiness. One without the other causes despair and frustration. For me, this book was an attempt to do something with the abundance of free time I had and still have between patients. I had to do SOMETHING with my free time besides watching YouTube and endlessly browsing Facebook or my Google News Feed. If I couldn't help others by fixing their health, I would share what I have learned through all the other experience I've had and knowledge I've gained in the process.

Not long ago, I watched an interview of Elon Musk, who made a fortune on Paypal, who founded SpaceX and Tesla, and other companies. He was asked, "How would you respond to those who want to be you?" He replied essentially, "I don't know that anyone should want to be me. I don't think being me is as great as they think."

In another interview, he described starting his companies as, "Eating Glass." Of course he has a comfortable life, but it's far from stress-free. Mo' money, mo' problems. We trade the problems of poverty for the problems of wealth. Some of us would rather have the problems of poverty, others would be terrified at the thought. They lose sleep and begin sweating just at mentioning it.

Now when I say, "Happy", I don't mean the short-lived kind of happiness you get when you happen to be eating delicious food, enjoying some romantic company, or getting a buzz off some substance or thrill. You know, that dopamine-oxytocin-glutamate-adrenaline rush of hormones.

Those things feel great, but their effects are very temporary and short-lived. Happiness is never permanent, but you can actually change the shape of your brain to be happier for longer and with less effort. A temporary rush of dopamine, glutamate, or oxytocin certainly feels good, but until you've built your life in such a way to keep those things flowing in a steady stream, they dry up real fast. Ultimately, our long-term happiness comes from an absence of constant stress, fear, and a steady flow of those hormones that boost our growth and progress.

Too often, we create an illusion of happiness when we keep giving ourselves a large shot of stimulation-based-happiness. We want a stream of these feel-good hormones flowing through our mind so we do what our brain needs to get that rush. Maybe it's a six-pack of beer every night, or daily masturbation. Perhaps it's 3 hours of TV every evening before bed. These rushes are short-lived, become a daily dependance, and grow to be slightly harder to achieve over time, even if just a tiny, tiny bit.

The first high someone gets from taking cocaine will often be their strongest because they are most sensitive to it. The brain sees this dump of dopamine and adrenaline and starts desensitizing those nerves because nothing should feel THAT good, and actually be good for us. So forever after, many addicts will keep trying to pursue those early highs and never able to achieve them, or they'll need more and more drugs to get it. Or sweeter and more sugary foods, or edgier and edgier TV and movies, or bigger and more dangerous risks.

Think of a river with a lake further up the valley. We love the rush of water and the abundant life it can bring to a farm when we flood the river with a bit of extra water. We get a nice abundant crop, a bit of excitement playing in the white-water rapids, and it's just pretty to look at, isn't it? But to get that extra rush of water, we have to open the flood-gates, or lower the level of the dam, or carve a deeper canyon at the exit of the lake. If we keep that kind of behavior up, we end up draining our lake. We've run dry.

To get that kind of water flow now, we have to wait for the lake to fill up, and while it's empty we have no reserve to get us through a hard time. We have no backup, no savings. If a drought hits us now, we are in serious trouble! And to get the lake to fill back up, we'll likely have to rebuild the dam, or canyon wall, or raise the flood gates, and live off of less water while we build that reserve again. In life, it's the same, we need a sustainable plan that can keep life as steady as possible. The world will throw enough curve balls and unexpected surprises at us without us making it worse. "Life's hard. It's even harder if you're stupid." Apparently, John Wayne DIDN'T really say that, but it sounds like something he would say. (It comes from George Higgins' book *The Friends of Eddie Coyle.*)

If we spend too much of our energy on pursuing excitement or trying to fill those empty holes in our life, we can drain ourselves of emotional reserves. We can suck our relationships dry, often ruining them. Then when we've completely run out of our emotional, hormonal, and social resources, we have to slug through the hard time rebuilding, restoring, and renewing them. And it's much harder to get through that process than to just never start it. Learning this fact of life is something that each person will have to learn in their own way, and we will all go through that learning process at least once at some point in our life.

Sustainable happiness is achievable, at least for periods of time when our environment around us is stable, and even many times when it's not stable. We can learn to be happy even when everything around us just plain sucks. And if we've learned to achieve that sustainability, we can rebuild our life much faster and easier when our environment around us (such as our relationships, our jobs, our health, the weather, etc) does fall apart, because it probably will at some point.

The first thing about happiness that just about every expert in the field agrees on is to be GRATEFUL. Be THANKFUL. Be OPTIMISTIC. These are all the same. Look around you and see what you have. Start with that, no matter how little it actually is. Whether you're homeless, hiding from the storm under a bridge in Kansas, or a billionaire in Dubai, if you want to find happiness, start with what you already HAVE.

Be thankful for the bridge. Be thankful for the 10 Lamborghini's in your garage. Be thankful for the clothes you have, the food you are getting to eat, the heart beating in your chest. Be thankful for the air you are getting to breathe (even if it is polluted! Be happy that you're not suffocating!) As I've visited support groups, as I've met with experts and listened to lectures on this topic, as I've attended worship services of my own and other faiths, as I've studied HISTORY, I've learned, it can ALWAYS be worse. Always. Be happy with what you have. And just because you seem to have it all, don't assume that you actually do!

Be grateful for your family and friends. If you don't have any friends, be grateful for those people who serve you at the grocery store and smile at you. Be grateful that most of the people around you are obeying the laws of traffic and driving in their lanes, and not running into you. Be grateful for challenges that teach you what you don't already know well enough. Be grateful for those purposes and drives that are fueling your need to get up in the morning.

Be grateful for your FAILURES. Recognize them as the only way to get better. They are the stepping stones to success. Be grateful you recognize them as failures! The truly unfortunate people in this world, are those who don't know that they are messing up. Self-awareness is huge. Be grateful for it.

There are those reading this who'll be scowling right now, or laughing their heads off thinking that being grateful for such pitiful things is just self-delusion. We do need to be careful that when we are counseling a friend or family member, deep in the pits of despair, we don't try to force that gratitude onto them. "At least you HAVE a family." Or, "Be grateful you've at least got your health."

Those admonitions, well-intentioned as they are, often come across as calloused or unsympathetic. Often, we do need to acknowledge and validate others' feelings before they'll feel safe enough to address them themselves. I can't and won't try to tell everyone what they should be grateful for in each circumstance. It's not my place, and I just don't know what each person is dealing with. What's important is you using your own mind to find things that are keeping you alive and learning and be appreciative of them.

Shawn Achor is a successful business consultant, Harvard graduate, and expert on happiness and positive psychology. Dan Gilbert is a renowned author, professor of psychology at Harvard, and an expert on happiness. Robert Waldinger is a psychiatrist and professor at Harvard's Medical School and an expert on happiness. All three of them would agree that such gratitude-based "self-delusion" is not a delusion at all.

Happiness, whether it comes from the joy one finds on eating a delicious can of Chef Boyardee, or walking around the block on a cold moonlit evening, or eating dinner with friends, or receiving a smile from your spouse or loved one, are all genuine forms of happiness. There's no self-delusion in regards to happiness. And they'll all agree that gratitude is at the root of all happiness. "Made-up" happiness still gets translated in the brain as REAL happiness. If you can find something to be happy about, you will be happy about it!

So unless there's something in the water at Harvard messing with the smart people there, it should be easy to believe them when all three of these experts will tell you to start by being thankful. Let's add to their findings with the recommendation in Stephen Covey's "The 7 Habits of Highly Effective People" to quit complaining and take charge of your situation, whatever it is. Use what you already have, or show appreciation or gratitude for what you've got, then go use it to the best of your ability! This is a hard logic to refute. Enjoy what you have. Work with what you have. Use what you have, no matter how little or how much of it you have. This is the start of happiness, and it's the start of change and success.

A popular habit of happy people and those who'll teach us how to be happy is to make lists of what we are thankful for. Do it on a daily basis. Whether it's a prayer to a God thanking him for your home and food, thanking the people you meet for all the good you encountered during the day, a simple list in a diary, or a quiet meditative moment before bed or right after waking up. Reviewing and reflecting on the good things that happened during a day set the hormones in your brain going in a good direction.

For the next several hours, or even the next couple of days, your brain will be working better: You'll think more clearly, more constructively, more gratefully, and more optimistically. You'll find answers to problems you had been unable to solve. You'll talk to people in more civil ways that'll be more persuasive and constructive. You'll find more success in your efforts.

Shawn Achor and his colleagues will tell you that MOST of the time, happiness MAKES you successful, not the other way around as most of this world believes. Most of our society is set up to think that success leads to happiness. It's not. Being happy leads to more success. Maybe not immediately, or as much as you'd like, but it starts, just like a tiny seed, soaking up moisture in the ground, getting ready to sprout.

The moral of this chapter is: Be Grateful.

15

Get Comfortable Getting Dirty but Try to Stay Clean

As I've worked with recovering addicts and visited support groups, I've seen those who are optimistic about their recovery, who don't beat themselves up for slip ups. Those who look at each new day as a challenge to overcome are those who make the greatest progress. That's not to say they don't recognize their failures, but they look at those failures with, what's called by most scientists, teachers, and experts, as a "growth mindset." They look at each failure as a stepping stone, something to learn from, an experience to be grateful for. In this way, they and their brain look for solutions, rather than collapsing under the despair, discouragement, or anger of another mess up.

A Growth Mindset is the belief that you can change, that you can become better. That your brain and your mind can grow. It's real, whether you believe it or not. Students who believe in a growth mindset do better on tests, perform better in classes, and suffer less stress and anxiety over the topics and subjects they struggle with. Adults and kids who believe in a growth mindset are able to learn new skills faster, form new habits easier, and tackle challenges with more success.

Along with having a growth mindset, the idea that you can change, that you can get better, comes the idea of practice and learning. Old Dogs can still learn new tricks. If you're still breathing, your mind can change. When you're a kid, a child, or even a teenager, just about everything is new and learning is just a part of being awake. When you've become established in a career, a lifestyle, and family for a few decades, life becomes routine and not much changes. Learning becomes uncomfortable, change becomes harder. We just aren't used to it. That's all fixable, though.

It's not that our brain can't learn anything new, it's that we forget what it's like to have to learn new things. Learning is hard. Practicing is tough. Failing and messing up, trying and trying and trying again is frustrating and painful, mentally and emotionally, not just physically. Just accept that and move on with whatever you're trying to accomplish. Maybe it is harder for an adult to learn a new language, but it's still very possible and happens all the time. Try to remember how hard it was to learn as a kid when every experience was new. My kids took a few years to get a basic grasp on English, even when speaking it every day. Most adults can pick up a new language in about a year in the same circumstances.

Read and Learn. Don't ever assume you know *enough*. Don't ever think that instinct or nature or biology will teach you what you need to know by itself. Through practicing and mistakes you may learn much of what you need to know, but you don't need to learn it all from scratch yourself. Take advantage of others' struggles and learn from them. Most of us have heard that "A Wise man learns from his mistakes." A few of us have heard the additional, "An even wiser man learns from the mistakes of others."

Having this growth mindset requires an open-minded progressive mentality. Don't hang onto everything you were taught before. You need to be able to look at new things, new ideas. You need to consider that there's still a lot more to learn and that others out there have figured out what you haven't. This doesn't mean abandoning what has worked for you your whole life, but accepting the idea that not everything HAS worked for you your whole life and you need to do something about it. It requires an open mind to possibilities. To new ways of thinking and living... at least in part.

I'm not saying ditch your home in the country and go live in the city near the universities and look to them for all the answers. Nor am I saying ditch your career, home, and family to go discover yourself wandering the world alone for five years. I'm saying, read books, look to experts, take classes. Focus on one problem or thing at a time, and figure out how to do it better. If you can't solve a particular problem, put it on hold. Don't abandon it, just set it aside and expect that you'll get an answer for it someday. While you wait, focus on other problems and solutions. Go out and get some answers!

Now, use all the answers you find with a grain of salt. There are lots of people who will give you answers freely, and lots more who'll love to charge you an arm and a leg for their answers. There are those who will tell you exactly what you want to hear, and those who'll love to catch your attention by spewing out the exact opposite of what you've ever been told. And some people will tell you what you already know but rephrased in a way that sounds all wrong to you. Not all these answers will be right, in fact, most of them will be wrong. It will take practice, experimenting, and even failing, along the way. Over time, you will become your own best judge of what is right and wrong.

Michael Pollan, in his book, "Second Nature" gives a great description of what makes for a "Green Thumb" or a person who is very good at growing plants. Even if you don't like gardening and have no interest in it, you'll find yourself laughing and thinking deeply as you read through his entertaining stories about it. He says it takes a great deal of practice and failure to become a green thumb. My experience with failing to grow my plants would agree. His opinion is that an expert in gardening is someone who doesn't get discouraged about failure, but sees it as a chance to learn something new about their garden, their weather, their soil, their environment, their habits, whatever it is. After enough accumulated failures, Green Thumbs develop an instinct for what works and what doesn't. This "intuition" ends up making gardening look easy to those on the outside or to the noobs who're still figuring it out.

The famous quote from Teddy Roosevelt, known as, "The Man in the Arena" has been an inspiration to millions of people. It's used by the military and public and private corporations alike to instill the drive to go out and do —to go and try.

"It is not the critic who counts; not the man who points out how the strong man stumbles, or where the doer of deeds could have done them better. The credit belongs to the man who is actually in the arena, whose face is marred by dust and sweat and blood; who strives valiantly; who errs, who comes short again and again, because there is no effort without error and shortcoming; but who does actually strive to do the deeds; who knows great enthusiasms, the great devotions, who spends himself in a worthy cause; who at best knows in the end the triumph of high achievement, and who at the worst, if he fails, at least fails while daring greatly, so that his place shall never be with those cold and timid souls who neither know victory nor defeat."

Leo Tolstoy, the famous Russian author, wrote, "It is true that I have not fulfilled one thousandth part of them [the precepts I profess], and I am ashamed of this, but I have failed to fulfill them not because I did not wish to, but because I was unable to. Teach me how to escape from the net of temptations that surrounds me, help me and I will fulfill them; even without help I wish and hope to fulfill them. Attack me, I do this myself, but attack me rather than the path I follow and which I point out to anyone who asks me where I think it lies. If I know the way home and am walking along it drunkenly, is it any less the right way because I am staggering from side to side?"

Failing and trying again is the key to success. And failing while learning from others, is the key to the greatest success. There's a saying that goes, "A master has failed more times than the apprentice has even attempted." I'm gonna repeat that again, later.

This idea has been spread by philosophers since ancient times. Confuscious said, "Better a diamond with a flaw, than a pebble without." Shakespeare said in his play, King Lear, "Striving to better, oft we mar what is well." Voltaire, the French author repeated it famously as, "The Best is the Enemy of Good." This idea is closely related to the Nirvana Fallacy, where we attack something as all bad because it is imperfect. We do this to our political leaders. We do this to our religious leaders and churches. We do this to our parents. It is a black and white way of thinking with no room for shades of gray and color in between. We cannot find perfection in this life. We will aim for perfection and find things that are good, and even things that are better and best. And one person's best may be better or worse than another's. That's just how it's going to be sometimes.

Perfection as an ideal is a wonderful thing. No one likes making mistakes. They hurt and are painful. However, if we don't act, hoping to achieve perfection first, we'll never act. Some people call it Paralysis By Analysis. We overthink a problem and never tackle it. More often, Good that is done is better than a Perfection that can't be achieved. Some leaders in the Armed Forces of the American Military call it the 70% rule. Basically it states, "If you feel 70% certain about it, do it. Most of the time, you'll be right."

The reality of life is that we very often can't see the entire pathway to perfection. We can't even see more than a few steps. So we walk those few steps, encounter a bend in the path or an obstacle, and then we make a plan to overcome it. It may take some time and slow down our progress. The goal is to keep making progress through our mistakes. We cannot grow without messing up. A child cannot learn to walk without falling down. If you are going to fail, fail quickly and learn from it.

For a long time, I tried to avoid failure in my life by researching and learning and analyzing the mistakes of others, trying to avoid failure at all costs in order to be the "wiser man". In reality, two things happened, I either didn't recognize the failures I was already making, or I didn't grow and progress for fear of failing. I was my own limit. I was my own curse. I had damned myself like a dam stops water. In a sense, the religious use of the word "damn" is the same. Your progress is halted. Your road towards perfection is hampered. When you "sin" or "transgress" the laws of God, or nature, or evolution, you are stopping yourself and those around you from moving forward.

Using both our own mistakes and the mistakes of others will make us wisest of all. Personal effort and learning from others' makes for the greatest chance of success and happiness. It allows us to do the impossible...or...at least what has never been done before.

By the 1940's a handful of athletes were determined to run a mile in under 4 minutes. Roger Bannister, a medical student, was one of those. For years he and others across Europe and America were trying to beat that milestone. There were many who thought it couldn't be done. Roger Bannister was the first beat that limit in 1954, only after spending a significant chunk of his life and free-time working at it. He only held that world record for 46 days before others were beating it. Many, many, great athletes now beat that four-minute barrier regularly. As of now, even at least 8 high school students have beat that record and two men over 40 have beat it. Currently, the world record for the mile run is 3:43.13 seconds.

Often we don't think something can be done until we see someone else do it. Once it's out that something is possible, others will follow. After you finish this book, go read about longevity and centenarians, and the feats that many of them accomplish. It's amazing! Anyone who competes in running can tell you stories of men who don't begin running until they are 80 and then go on to complete marathons. Or see how many 90 year olds who go back to college and earn bachelor's degrees. There are scores of folks who retire from a full career and start new ones, from scratch. As I learn what the very elderly can accomplish, it makes me determined to be one of them.

I understand that physical and mental limitations are very real for some. Some of us just will never be able to accomplish certain things, that is true. BUT (that's another big BUT) we don't know what we can or can't accomplish if we quit trying, or never even bother to start in the first place. If we want to be the first to do something, expect it to be nearly impossible, because, if it were easy, other people would already be doing it! Everyone thinks it's impossible, because no one else has figured out how to do it yet.

A friend and colleague of mine showed me a trick to practice doing oral surgery around sinuses, since we have to do that once in a while as dentists. He would practice using raw eggs by removing the shell of the egg from the inner membrane without tearing it or ripping it. He posed a challenge to me to remove the entire shell from the egg. He didn't think it could be done and told me of the troubles he ran into as he tried. Searching online couldn't produce any results of others getting further than he had. Even a news article of a neurosurgeon from China who practiced his hand skills by doing this only showed him removing a small portion of the shell. I learned from my friend's troubles, came up with a solution, and after 9 tries, I was able to remove the entire shell, a tiny bit at a time. I even got it on video and put it up on Youtube to brag about it to my friends and family. I was so proud, can you blame me?

Arthur Boorman is a former military paratrooper, injured from his military service. He became severely handicapped and was told by doctors that he would never walk again without the help of someone or something else holding him upright. For many years he lived with that "reality" relying on crutches and leg braces to get around. He had accepted it as fact for many years that he was handicapped, crippled, stuck. Until something in him decided he would find some way to get better. He found a brave Yoga teacher with a unique program including a strict diet and careful yoga sessions. Arthur slowly learned to stand and balance on his own without the help of crutches. Then he learned to walk. Eventually, he learned to run! Without the help of crutches, braces, or aids! He gained a whole new life because of his own determination and the help of other brave people.

The lesson is that we shouldn't give up. The greatest limitation in our life is the limitation we put on ourselves. I always did like the song "The Curse" by Michael Mclean with the lyrics, "There is no curse or evil spell that's worse than one we give ourselves."

When we struggle to overcome some sin or weakness or addiction in life, we can take two approaches. 1. Believe that one day we can become better from it. That God, a higher power, or the World around us, will one day help us overcome it. We can believe that struggling with such a challenge is just a normal part of life and actually improves other aspects of our life. Even if we don't win in this one area, we often learn empathy and kindness to others in the process and thus improve one one of the most useful skills of life: Building Relationships.

The second approach to dealing with our weaknesses is 2., After spending years or decades fighting against something we have little to no control over, we quit, give up, and just begin openly indulging in whatever we feel like, since it doesn't make any difference anyway. "If I'm gonna keep failing, I may as well jump all in and enjoy it!" That's a fast way to lose everything you love within a short time period or get yourself really stuck in a bad habit.

Even trying to decide between these choices is a normal struggle. Some days we will want to just quit and give up. Some days, nothing will stand in our way. That's normal. Having a growth mindset —the deep belief that people can change, that we can get better, become happier, and improve ourselves— is crucial to finding happiness and riding out those low spots when we feel like quitting.

This idea that things can get better is deeply rooted in our idea of Love. When we find things to be grateful for, we appreciate them, we sacrifice for them and pay prices for them. We begin to love these things. We begin to give true value to them, not just financial or economic value, but emotional value. That value that the Mastercard commercials tells you is "priceless."

When we realize that we can change, we begin to love ourselves, not in a narcissistic way, but in the way that you love a baby child. We see ourselves with potential the way our loved ones see us. When we love ourselves we begin to love others. We slow down our criticisms and judgements of others, understanding they are just like we are, weak, flawed, and still learning, still growing. We learn that many of our personality traits we think are "natural" aren't set in stone and that this applies to others. We become just as hopeful for others' growth as for our own, and we begin to trust in them to work on themselves as we work on ourselves.

Practicing gratitude and a growth-mindset takes time and practice. It will not come easy all the time, and it will not be quick, or without major setbacks at times. Like Jim Collins reports in his book, "Good to Great," on his research of companies who excelled, while others in their field with all the same advantages failed: success and improvement is like turning a great big flywheel. You have to generate momentum a little bit at a time, but a small force —applied consistently and regularly— creates momentum. One day, those around you, including yourself, will look and see what you have accomplished and say, "Wow" how'd you do that?!" When they do, tell them, "Magic!" Just kidding. Tell them the truth.

The take-home message: Take the first step.

16

You Are Not Stuck, It Just Seems Like It

In the book "Influencer," authors and researchers Joseph Grenny, Kerry Patterson, David Maxfield, Ron McMillan, and Al Switzler, do a great job outlining all the different factors that lead to change. To give an overly brief summary of their incredible book, change is brought about by a mix of Ability and Motivation and driven by personal factors, social factors, and environmental factors. For personal, social, or environmental reasons we feel motivation to change, or we may be aided or hindered by the nature of those things.

These ideas are scattered through other philosophies, but their book brought it all together, into a nice orderly pattern. Their book gives the reader knowledge and ideas of how to improve the behavior of ourselves and others.

If you've read it, or read any other self-improvement books, these ideas should be familiar. They are pretty universal. If you're familiar with Robert Cialdini's "6 Principles of Influence," you'll find that his 6 principles fit into this philosophy nicely. Like I talked about in the last chapter, we need to get over the idea that we are stuck with our genes, stuck with our lifestyle, stuck with our talents. We are not. Malcolm Gladwell wrote in his book Outliers, that the difference between the most "successful" people in life and the rest of us, DID have something to do with opportunity, privilege, and birth, but the other half of the equation came down to effort, hard work, and determination.

Bill Gates had been coding and programming so long and from such an early age that by the time he was a young adult, he was leaps and bounds ahead of his colleagues. The Beatles had a tremendous amount of time together as a band playing clubs in Hamburg. Enough so that by the time they started attracting media attention, they had a fairly well-polished act and group together. Wolfgang Mozart's father was a well-known composer who helped his son begin composing at a very young age, so that by the time most of us are entering college, Wolfgang was performing for royalty.

We still see this at play in our lives. Michael Jackson was performing non-stop with his brothers since he had been a young kid singing "Rockin Robin" on TV. By the time he had reached adulthood, he was a legend. Some people just have opportunities that will propel them to fame and notoriety, especially when amplified by ridiculous amounts of hard work. Grant Cardone is a huge promoter of this idea in his book "The 10x Rule." Just about every setback can be overcome by huge amounts of work and effort. Opportunities are often MADE, by us.

But those opportunities don't all look the same. Oftentimes, our handicaps become our greatest accelerators to success. In another of Malcolm Gladwell's books, "David and Goliath" he gives story after story of those who were given awful situations in life and how those awful situations actually became the pathway to success and fame and skill. Often, our setbacks become the way to greatness. Our weaknesses are actually our greatest strengths! He shows that by most opinions, losing a parent while still a child can be one of the greatest setbacks a person can face, yet, a largely disproportionate number of the Presidents of the United States of America have had to deal with that challenge as a kid or teenager.

Being Left-Handed has a number of drawbacks, most not very serious. Before the 20th century, left-handedness was seen as a disability and children were forced to use their right hands, despite their preference. A forced-conversion therapy that left many emotional scars was prevalent. All major world languages are formed to exalt right-handedness as the correct way of being, with left-handedness as being "sinister," "crooked," "wrong," or "false." In English, when we say something is "Right" as in "Correct," it literally refers to right-handedness. Cursive writing was designed to be done by a right hand, as just about all lefties could tell you. Many power tools are designed for right-handers. Try to use them with your opposite hand and watch out. A circular saw that you can't easily turn on and off with your opposite thumb is a little sketchy in your left hand! In the kitchen, measuring cups and all sorts of cooking equipment have spouts, handles, or measurements favoring righties. Most large measuring cups in the USA have the metric measurements facing you when you pick it up with your left hand.

When I was in dental school, I noticed almost all of my class's lefties were assigned to the extra remedial practice sessions for the simulation lab, where we practiced working on plastic teeth on the mannequins. I pointed it out to one of our faculty and we almost wrote an academic article on it. Our hypothesis was going to be that the early assignments were geared to be easier to right-handed students, but unintentionally harder for the lefties. The later assignments that were more of a challenge for right-handed students were finally easier for the lefties.

Being left-handed wasn't too much of a problem as a teenager or young adult, but as a dentist, it can be a big hassle to equip an office for me. Mostly it just causes lots of funny quirks. When I play card-games, I often hold and fan out my cards upside down. "Why do you do that?" people always ask. The cards have their numbers in the upper left and lower right corners to be easily seen by the holder when fanned slightly. I have to make one of two concessions. I can either hold my cards right handed like everyone else, or I can fan them out upside down and just read the numbers upside down. As a kid, I guess it was just easier to do the second, and that's what I've always done. Go figure.

Even though only 1 out of 10 people is left handed, 6 of the past 12 Presidents have been left-handed, and many of the runner-ups who lost the Presidential elections, were also left-handed. The thinking is that left-handedness doesn't offer many inherent skills or abilities, but the ability to learn to adapt to a right-handed world changes the mindset and problem-solving strategies of lefties. Lefties tend to be more creative, and adaptable. Studies hint strongly that these are skills which they learn at a young age to cope with all the little (and sometimes big) things in life that are set against them.

While I was working toward a career in Civil Engineering, I fell while hiking and sliced the tendons to the ring finger of my left hand. A few months of wearing a splint, completely immobilized during the healing, caused my hand to be nearly useless. Several months of physical therapy were prescribed. By the time the next semester was starting, I still couldn't write with that hand, so I started practicing with my right hand. I had already finished my summer internship by learning to type one-handed. All those years of piano lessons gave some head start on learning to jump around a keyboard for my right hand. Writing was another obstacle entirely. It was frustrating and maddening. I had to choke back tears a few times those first two weeks of school, angry and worried about how to take notes and keep up in class.

I did make progress. I could write legibly, even if slowly. The progress was real and tangible, but not fast enough for me at the time. In those circumstances, it became easier to accelerate my physical therapy and get function back to my left hand, so that's what I did. I know now, that if needed, I can learn to just not be left-handed if push came to shove, but it was easier if I didn't have to change in that way, at that time. So I didn't.

Oddly enough, going through the surgery and therapy from that injury is also what awoke a fascination with anatomy and the human body and the desire to go into medicine of some kind. I have full functional use of my left hand again. The one finger is a little weak, but it doesn't stop me from much except playing bar chords on the guitar. (I guess I could learn to string my guitar lefty-style like Paul McCartney, but I'd rather learn to fit in with other guitar players, especially since, I don't play much anyway). What began as a weakness changed my life to pursue something I didn't realize I would have enjoyed, and taught me that, even though I didn't have to change myself that time, I knew I could for other things. I may have always been naturally left-handed even if I'd learned to become right-handed to the world. I've met a few dentists and heard of more, who either forced themselves to learn to practice dentistry right-handed, or injured their left hand and were forced to do it, and still others practice using a weird combination of both hands to accomodate the offices they work in.

We like to tell the story of the little shepherd, David, against the giant war-hero, Goliath. Interestingly, the apparent "weakness" of the biblical David and his little sling was the greatest advantage he could have had over a large, powerful-yet-sluggish, slow, weighed-down Goliath. A sling in the hands of a skilled user can be as or more powerful than a handgun. Match that against a lumbering giant, trying to fight with a shield and sword and it's no match at all. Like bringing a gun to a knife-fight.

We don't know what we are capable of until we try. My kids get to hear me say all the time that the words, "I can't" are not allowed in our house unless it's followed by a "yet", as in, "I can't yet." "I don't know...yet" is also an acceptable answer, because then we can focus on what they need to learn or practice.

Remember from last chapter? It was Stephen McCranie who said, "The Master has failed more times than the beginner (or apprentice) has even tried." Failure is a refining fire. It purifies us. I don't mean intentional failure where we go commit a crime on purpose just to "learn from our mistakes", that's like feeding our hand to a shark and calling it "character building." I'm talking about the kind of failure that comes from trying to improve. Failure, messing up, coming short and recognizing it, is how we change.

Too often, most of us aren't willing to change ourselves. We aren't willing to correct or alter our persona. We are fine just the way we are. "Are you telling me I'm not nice enough???" (said in an accusatory or threatening tone....) What about being true to yourself?

The idea of being "true to ourselves" is a strange one. We all are constantly in the process of becoming someone or something else, whether intentionally or not. Stormwater will run downhill every time, but we can steer and guide it into storm drains and flood channels so it doesn't end up in our homes, most of the time. We can save it and store it and sometimes even treat it and make good use of it instead of just letting it destroy what we work hard to build.

We will all tend towards certain behaviors, some good, and some bad, but we can steer them into productive, or at least, less destructive places. And if you can steer your tendencies into a less destructive direction, a little at a time, you may one day be able to steer it into a positive or constructive direction.

What we spend our time and thoughts on will determine what that something else will be. If we spend it on pornography, we are becoming a pornographer. If we spend it on gardening we are becoming a gardener. Usually we're a unique mix of common things, a cyclist-CEO-husband-father who surfs on the side. Something like that. It will change throughout your life depending on where you are, what you're working on, and what you have to work with.

Don't worry about being true to yourself. You can't really know who your true self is. You either decide what you are going to work toward, or you will be the Self that the world around you makes you into. Your identity is fluid and dynamic. Your sense of humor and style might not be quite so visibly fluid, but it is, even if it takes years of slow change, or some major traumatic event. As we learn more, we will grow. As we try, we will change. As we want, and think, we will become. As "life" hammers away at us, we will slowly change shape. Our hormones change, our environment changes, our families change. You will not be the same person in five years that you were ten years ago. Sure there will be many, many similarities, but you will be different.

Cal Newport, in his book, "So Good They Can't Ignore You" discusses the need so many of us have for pursuing our "passions." He gives several examples of celebrities and leaders who promote this "Passion Mindset," this need to pursue the career or lifestyle that stimulates and lights up our mind, like a drug. The problem with the Passion Mindset is that it focuses on what the world, or our career, or our pastime can do for us. What do I get out of my job, what stimulation and excitement does it provide for me??

This is a selfish way of thinking and causes a great deal of anxiety, torment, and broken relationships for those who pursue it. If all we chase is our Passion, often we end up losing our material goods, our freedom, or even our passion in general. Then we go chasing some other passion and the cycle starts over. We never get to earn the privileges that come with higher responsibilities, or the challenges that come with expertise. And interestingly, many of these ultra-successful people who preach the Passion Mindset started their own careers by not following their passions, by simply pursuing a job that they ended up getting really good at.

The opposite is the "Craftsman Mindset." This is when we let our skill determine our love for our job. Just like an apprentice becomes a novice, who then becomes an expert, and finally a master. When we think of a true Master at something, we think of a person who eats, lives, and breathes their craft. No one is born a Master. It takes years of incredible work, effort, and dedication. The Craftsman Mindset focuses on "What can I do for others that is useful or valuable or needed?" It is unselfish by nature. It teaches us to focus on what others need and want, and how can we go about getting it to them. It is a never ending pursuit. As we work towards mastery we gain more responsibility, more freedom, and bigger challenges.

We run into trouble when two things that we want conflict with each other. If we want to drink and party constantly while holding down a job and family, chances are there'll be some conflict and we have to give up one if we want the other. We can't always have it all. Sometimes, we begin to start trying to serve two "masters" or two opposing desires. We will either love one and hate the other, or hold to one while despising the other. If you're careful and set precise rules and boundaries, sometimes we can meet our wants and desires by putting careful constraints on them. It takes practice, supervision, support from others, and a lot of experience from others and yourself to get it right without destroying yourself or those around you.

You do have to pick and you do have to prioritize what you want most. And whatever you pick, you'll become that more and more, if that's what you spend your time and energy doing. We will have certain personality traits that we just can't manage to ignore no matter how many times we've chosen away from it. That may be true for several decades of your life. There are certain aspects that will never go away. I may just always love a good (or bad) pun, no matter how many times my wife tells me I shouldn't. I may always have an urge or tendency towards a temper, no matter how much of my life I spend reciting, "Serenity Now!"

This doesn't mean my life is in shambles or ruins because I can't indulge in that one thing, or things. If I become anxious about missing out on that "thing" I start to become consumed with it. Oftentimes we become consumed with getting the one thing we can't have. We become obsessed with what we don't have or what we think others aren't "letting" us get. It becomes our focus.

I've heard people call this the "Missing Tile" syndrome. When you have a floor or ceiling that is otherwise perfectly filled with tiles, except for a hole where one is missing, it's easy to ignore everything that is right, and focus on the one gaping flaw. We do that with our relationships. We focus on the other person's flaws more than their strengths. We do it to ourselves. We fail to be grateful for what we have, and overemphasize what is wrong. We do it with our churches, our schools, our communities, our nation, and the world in general. We focus on the negative. It's easy and it feels good.

There's a common motivational quote thrown around large business settings and by life coaches and consultants of all kinds, "If you plan to fail, you'll definitely succeed."

It reminds me of this story that happened when I was going to college. It was a city park with lots of hills perfect for sledding after a good snowfall. Two boys were going up and down a particular run. Near the bottom of this run were some posts, for an old no-longer-existent gate, deeply embedded in the ground, unmovable. Dozens of adults and kids had gone up and down this hill without any issue, zooming between these posts without any trouble. Witnesses watched these two particular boys start their run sharing a sled. The boy in front who had some steering ability, attached his focus on one of those posts, and with terror on his face veered his sled from its straight course right into one of the posts in an awful crash.

If you plan to fail, you'll usually succeed.

How often do we do that to ourselves? Now, in all fairness, I don't know that the boy actually steered into the post, or whether he had any control over it. But we've all seen ourselves or those around us get our focus stuck on the one thing we don't want and go careening right into it, head first.

How many times do we try to fix our spouse, or change their bad habits, and then wonder why they don't like spending any time with us? How many times have you seen the strictest parents have the most ill-behaved kids? How many times have you tried to force trust into a relationship only to have it backfire into betrayal? Becoming obsessed with what we DON'T want, is a bad idea. We need to focus on what we do want, and we need to look beyond that goal to make sure that what we want gets us to the type of end of life we want.

When teaching my kids to ride a bike, their tendency is to look down at their feet or their hands. They are trying to learn to steer and pedal at the same time. What happens instead is that without the far away focus, they wobble all over the place, they can't ride in a straight line, and so they can't learn to balance properly. They fall over constantly.

When they learn to look out ahead of them and focus on the farther direction they want to aim for, they become more stable, more steady. When their focus is on the immediate area around them, it's too late. I remember getting advice from a Driver's Training Instructor while learning to drive as a teenager. Many of his students would weave and swerve within their lane lines, worried about staying in their lane. When he taught them to look further down the road, they began driving straight, only making the minor corrections needed to maintain it.

If our focus is only on the boundaries and rules we have to keep, our lives will be all over the place. Constantly bounding and bouncing from one rule to the next, always frustrated by the "limitations" that society, or family, or religion has placed on us. Always keeping us "down." If we set the course of our life on a far away target, it pulls us toward it like a magnet and those boundaries and rules and commandments act only as insurance, or safeties, or guardrails, hardly noticeable until we gently graze up against it. Like a rumble strip on the side of the Interstate Freeway. A gentle wake-up that we lost our focus for just a moment.

Sometimes we steer our life to meet some biological or hormonal need, sometimes to fill an emotional void. Sometimes that's just the course our life got started on and we don't know how to steer away from the impending destruction. That doesn't mean we're stuck, it means, we need to learn more, and work at changing things. Sometimes, we just have to avoid the temptation for long enough that the temptation slowly —ever so slowly— eventually fades into nothing. The only way to change desires is not through will-power and sheer white-knuckled determination, it's through learning more, growing, and progressing. I used to think Pizza and Ice Cream for breakfast would be awesome. Now? Not so much. I used to hate getting served Oatmeal and hot cereal. Now, it's my favorite. Why? I grew up and with that growing up, I learned and tried all sorts of breakfast foods and figured out which ones make me feel the best throughout the day.

John Nash won a Nobel prize in Economic Science and the Abel prize for mathematics. He was portrayed in the 2002 movie, "A Beautiful Mind." In the film you see him struggle in dealing with delusions and hallucinations, a result of his paranoid schizophrenia. He interacts with characters and people that aren't really there, but whose existence —in his mind— allow him to indulge in feelings of importance and stimulation. Eventually he learns to avoid and resist these hallucinations, though they never go away. As it is portrayed, while he is being checked out by the Nobel Prize committee to see if his mental health would be an issue in receiving the award. He tells them, "I still see things that are not here. I just choose not to acknowledge them. Like a diet of the mind. I do not indulge certain appetites, like my appetite for patterns."

A popular Youtube channel put out a video condemning "Nice Guys" and how so many of the good guys out in the world suppress their sexual impulses to appear nice and gentle to the women they encounter, but then release those impulses in vulgar ways in private. The video condemned the repressing or holding in of our urges, hinting that doing so always leads to inappropriate or destructive behavior. I started reading the comment responses to the video and enjoyed seeing the first comment with almost a thousand "thumbs ups" to it. It said something like, "Everyone is suppressing something, that's how we're civilized."

Often times, when we feel like we are having to give up some vital part of ourselves just to "pay the bills" or "Keep up appearances," we are feeling like we are being robbed of happiness by others. We feel like we are denying some vital or necessary part of ourselves. This can lead us into depression, anxiety, or all sorts of feelings of "numbness". We have become a victim of our biology and our society being at odds with each other, with our conscious, individual selves stuck in the middle.

The truth is, EVERYONE has to deal with this same struggle. Everyone. Whether they realize it or not, whether they have anxiety over it or not, everyone has to struggle with indulging in biological desires versus meeting society's expectations. This is nothing new and nothing abnormal. Letting ourselves worry about this struggle, only makes the struggle worse. Much like how when we can't fall asleep, worrying about it only keeps us awake longer. For all of history, humans have had to learn to balance their individuality with their society. Sometimes society wins out, sometimes individuality wins out.

We fall into the trap of blaming others or our circumstances for the emotional troubles we're dealing with inside ourselves.

Too often, we think that if someone doesn't love us for who we are, they don't love us, or didn't ever TRULY love us.

I personally don't know anyone who's ever TRULY loved anyone else. I do know people who've gotten close. True love is a never-ending process, not an achievement. We are all imperfect and mortal. We are all flawed and weak. We are all selfish. One of my own biggest pet peeves is when I hear someone criticize another person's character because they are "selfish." That's like saying, "They are breathing air." We are all selfish to some degree. Yes, some more than others, but we are all selfish. We are all trying to please at least some of the people around us, and we are all trying to please ourselves. We can't depend on others' approval for our own happiness, but we can't ignore it either. We do have to live with others while maintaining our individuality.

Dale Carnegie has told millions of people, "The only way to keep happiness is to share it!" He also was a big proponent of hard work, teaching his clients and students to work so hard they won't have time to worry about whether they are happy or not." My dad has told myself and my siblings more times than I can count, "Work is the best therapy." I'll tweak that by saying, "Helping others is the best therapy." Which, usually involves work, of one kind or another.

We are all constantly being pushed and changed by our environment and circumstances. The person and people we are trying to love are constantly changing, and we ourselves are constantly changing. Love is a never-ending effort to stay attached to someone else. We cannot be true to ourselves, because almost no one actually knows who their true self is. They may know who they want to project onto others, they may know who they want to become, but a true self, is a myth, a dream. Loving someone else while you both change requires constant effort, like trying to care for a lawn and garden when trees and weeds keep trying to sprout up. Like keeping a home dusted and vacuumed while dust continuously settles. Relationships require steady work to keep.

Too often, the idea of being true to myself is me justifying some bad behavior that's hurting other people. You never hear the saintly and gentle among us shouting to each other, "Well, if they don't want to love me for who I am, then that's THEIR problem!" Or even worse, "If they can't handle me at my worst, they don't deserve me at my best!" In all honesty, it's usually the feisty, the troublesome, the selfish, the irritable, the angry, and the lazy that tout those kinds phrases. We all THINK those things from time to time, but most of us don't actually say it out loud, especially to others.

Another phrase that goes along with this is, "Good girls never made history." Actually, they did and do, all the time. While Helen Keller, Marie Curie, Mother Theresa and others may not have fit in with other women, they seemed to be decent and professional people while they fought for their causes. The kinda history that "bad girls" make, isn't USUALLY the kind of history we encourage. Yes, many women have faced tremendous adversity from society while making important changes and just about all of us agree that those changes made enormous, immeasurably valuable contributions to the world. Let's discourage looking for trouble and justifying it as making history. These women went looking to make history and, like every historical figure, found troubling obstacles to overcome.

What do we hear from the wisest among us when someone disagrees with them? Things like, "They don't have to like me. I'm okay with that." or, "That's too bad, what's up next?" We don't have to make everyone like us, but we shouldn't shove our worst selves right in their faces and then get mad at them when they refuse it.

We get to create ourselves. And what is it we want to create? Who do we want to be and why do we want to be it?

What are we motivated to become? Much of the time we are motivated to avoid. We are much better at staying away, than of becoming something more. We learn that hot stoves will burn us and we are motivated to NOT touch it again! We often learn from experience what NOT to do. Or we learn from someone caring enough to teach us that being grumpy, quiet, mean, or shy will eventually make us lonely. We may learn from experience that there are aspects of our personality that we aren't happy with, but don't know what to do about it. Maybe we want to change and become motivated to change, but don't know how!

Are we going to become just a default person that fits into our environment, or are we going to decide on an ideal and aim for that? You have to figure out why you are doing what you are doing. Are you just going with the flow? Or are you pursuing some cause? Are you looking for superficial pleasures? That may not be a bad thing, because as we grow and mature, we move past those superficial pleasures and become motivated by higher causes in the end.

Simon Sinek spent a whole book, "Start with Why," to illustrate how our motivations drive our behavior. He then wrote a few more books just talking about the "Why" we do what we do and how to figure that out, in business mostly, but also in our personal lives.

So then, what do we have the ability to create within ourselves? There may be tendencies that have been with us since we were very little that have been a part of ourselves for our entire living memory, but that doesn't mean they'll be with us forever, nor that you'll even want it for the rest of your life. Don't make decisions for the rest of your life because of a destructive habit you want now or have wanted for the past 20 years.

Sometimes, those destructive tendencies can feel overwhelming. They can take over. They keep getting in the way of us becoming who we want to become. They feel like they will rip us apart from the inside until we give in. Those tendencies shout at us in our heads that they will never give up. They will ache and gnaw at us until we give in. Our ability to resist just can't keep taking it. It can feel impossible to keep managing, but keep your options open for as long as you can. If there is one skill or trait that has helped humanity thrive on this planet, is it's Persistence and ability to Endure.

In a purely physical sense, there's a growing number of scientists convinced that it was not our speed, strength, or intelligence that allowed early, primitive, humanity to survive. It was our ability to physically out-endure anything we chased. Mentally, no other creature has solved as many problems or overcome as many obstacles as we have. No other creature has adapted to the various harsh environments of this planet as well as we have. No other creature has pursued the dirt and rocks around us with as much creativity to craft tools and technologies and the wonders of history as we have.

Endurance and persistence are what can save us more than anything. If you ever watch a predator stalk its prey, it can sit and wait and wait and wait until just the right moment to strike. It is it's persistence that allows it to eat and live. On the flip side, Cheetahs —despite being the fastest land animal— only catch about 1 out of every 20 gazelles they chase. What is it that saves all those gazelles? Their persistence to outlast the short sprints of the cheetah. They have a lot more on the line. The cheetah is simply looking for a meal. The gazelle is out to save it's life.

If you can't out-endure your demons, you may need find a way to let those things out safely. You might need an outlet and the environment in which to do it. Perhaps you can translate your demon into something else that satisfies it, or takes your mind off of it. Napoleon Hill spends a chapter of his classic book, "Think and Grow Rich," —written back in 1937— telling his audience that to be successful, one must "transmute" their sex drive into a drive for success. Commit to the things that build lasting satisfaction and service to others. Commit to being loving and hard-working. Don't stop learning, hoping, growing, and practicing. If nature or biology knocks you down, get back up and try again.

If you're dealing with a personal trait you wish you could get rid of, but haven't been able to, the worst thing to do is worry about it. Anxiety over your personality flaws will almost always just make those things worse. Understand that you are not facing any challenge that humanity hasn't dealt with before. Go find people who are going the same direction as you. Don't settle with those who want to sit around and complain to each other how hard it is living with the problem. Find those who are challenging their problems, confronting them and getting better at living with the challenge. Complaining and whining has a place. Venting and releasing steam is sometimes necessary to help us work out our problems, but it is a small, small place. Find people who will share with you their successes and their strategies for improving.

Are you worried about your eating habits? How about your physical attractiveness? Maybe it's a substance or behavior that you can't figure out how to stop. Maybe you've got a temper problem. Maybe, you've got a major problem that's triggered by a substance, or memory, or current experience, but you've never connected the two together. Perhaps your constant need to eat has something to do with the smells, scents, and thoughts of the bakery right below your office, even if you're disciplined and don't ever indulge in their sweet pastries! Your willpower gets run dry by resisting the temptation and you're more likely to give in to something else later.

Numerous studies have shown that if we've used our reserves of willpower resisting a temptation, our brain is much less capable of solving problems. For example, if test subjects are told to ignore the bowl of radishes in front of them for 15 minutes and then take a test, they do much better on that test than subjects who've had to resist the warm, fresh-baked cookies sitting in front of them before taking the test. Our brain gets tired fighting temptation. We need to learn how to keep our environment clear of as many temptations as we can if we want our brain working at peak capacity.

Maybe the fitness magazines you love to read are a form of porn for you and may be contributing to other problems. Perhaps the stress at your job or a grumpy boss or supervisor has something to do with the temper you can't manage to control at home, taking your frustrations from work out on your spouse or children.

Lots of times, when something we don't want is overwhelming us, it means fuel is being thrown on that fire by ourselves or our environment, and it's getting out of control. The rest of this book will give some ideas of how to cut off that fuel source, or calm that fire down, or even just accept the fire for what it is and use it for something constructive.

I'm not going to say that you should give up whatever it is. You'll know, or your family and friends will likely know better than me what you should give up. Listen to them and take their advice carefully.

Below are various ways that we get the message from others or from the Universe that what we're doing, how we are acting, or who we are becoming should change. Soon.

When we need to change but cannot on our own, we need help, from others, and from the world around us.

In a nutshell: Help is everywhere.

17

Social Motivation....or... It's Time We Talked....

This is basically when your friends and family gang up on you to convince you that you have a problem that needs addressing. Perhaps it'll be your coworkers. Maybe it'll be the police or a judge. It might be your community or neighbors. Maybe your boss is trying to tell you what you need to change so that you can keep working in your position. Maybe they're right, or maybe they can just go stuff their complaints in their pie-hole!

Maybe your little kids are just crazy-wild and bonkers and none of your friends ever had the guts to tell you until they all decided behind your back, that, yes, you need to step in and reign in the little ones a bit better. Perhaps you are finally getting told that your breath reeks and you need to stop storing your clothes in mothballs, because, being smelly isn't socially acceptable. Maybe your wife has told you for years you should quit drinking because you're an awful drunk, and now, your kids are telling you the same thing. This is information you need to carefully consider.

Sometimes it'll be a good friend. Sometimes a parent. Sometimes it's a formal meeting at your house. Sometimes it's a conversation over a cup of coffee. Sometimes it'll be a mentor. Sometimes it'll be an employer.

Whatever it is, love needs to be the primary motivator. If it's not, be careful.

Because the choices of an individual, or a small group can affect the whole group, Life (not just humans!) developed the use of shaming, shunning, and embarrassing those who behave poorly. It works, but it's also hard to put an appropriate limit to it, or use it correctly. It's easy to go too far using it. Or be too free in dispensing it. As Americans, we're trying to curtail it's usage, but we're terrible at it. We call out slut-shaming, or fat-shaming, or the like, but we publicly crucify anyone who mispeaks on twitter. A few men got caught "mansplaining" recently and the internet just about blew up. Shaming is still as prevalent as ever, maybe even more so now with social media and our interconnections. A young woman makes a poorly-chosen caption on Instagram for only friends and family to see and within hours hundreds of thousands of strangers can tell her how awful she is.

Love needs to be the motivator and the method.

But wait! This gives everyone license to criticize everything about anyone as long as we "love" them! Yes and no. Yes, because we are all supposed to help each other. No, because it needs to be done with tact, care, love, patience, and all that good stuff. Timing can be everything. Wording can be everything. In the book, Crucial Conversations, "Make It Safe" is the mantra. However you confront someone about something, do it in a way that they can still feel safe and unthreatened dealing with this new information that requires some change from them. It's a fantastic book with lots of other good ideas. Go get it. Now. Or... after you finish reading this book.

John Gottman's book "The Seven Principles of of Making a Marriage Work" is geared towards married couples, but there is some fantastic advice in there about discussing touchy, emotionally-charged subjects with anyone. When you have to tell someone something awkward, keep it simple, factual, and specific. Don't get side-tracked by distracting topics, or defensive attitudes, or insults thrown back at you. Give them a "way out" or something they can blame their problem on so that the other person isn't as likely to take it personally or get embarrassed.

For example, "You've been doing a great job here, we all love having you on our team. I don't think anyone ever mentioned to you the correct order for accomplishing this task. I wanted to talk to you about that and see if we can help you do this even better."

What if someone tells you that you need to shape up but they do it in less than a tactful way? What if they were jerks about it? What if they are wrong? Do we get off scot-free from having to fix ourselves just because they weren't careful about how they said it? Nope. Just because you got offended at having your imperfections shown to you by someone else, doesn't put the responsibility on them. You still need to change yourself if you need to change yourself.

And what if it's just their opinion? You would do well to listen carefully and at least give serious thought to the opinions of your family, spouse, friends, and even coworkers and yes, sometimes even strangers. It doesn't mean you have to do it, but you should at least consider it before dismissing it outright.

Interventions, crucial conversations, and honest opinions are part of the "Social Motivation" realm of things that can influence us. We feel motivated to change because others around us want us to.

Again, if you need to hold an intervention, do it in a way that makes the other person feel safe. Discuss only the facts, even if those facts are your feelings. "When you behave that way, it makes me feel disrespected." Be open to conversations, validate their feelings and make sure you can express your own without attacking their personality. Keep it focused on the outward problem, not their personality. Telling someone, "The reason why you spend so much money is because you're so pig-headed about everything" has never done much good. Instead, "when you bought that jet-ski, you spent money that we needed for medical bills. That throws off a lot of our other plans and makes things complicated." (just an example).

See what I did there? Focus on the facts and details, not the personality problems of the other person. When we blame another person's poor decisions on some basic personality flaw, it's called (officially) a Fundamental Attribution Error. We tell ourselves, that person did a bad thing because they are a bad person. We all make mistakes even though we are all, basically, and usually, good people. We need to avoid the temptation to think that "those" people are worse than us. This is easier said than done, and for those who master it, life is much less stressful.

Whichever side of the conversation you are on, do your best to keep it calm. If you feel yourself starting to get excited, that's your "Go System" taking over —your fight and flight adrenaline-pumped-system— and your thinking will start to suffer. You're getting flooded with hormones and emotions. Re-focus your attention on the task and goal, not on your emotions, or the other person's. If you need to, ask for a break before you continue the conversation, or look for a way to lighten the mood just a bit, to de-escalate the tension. Work an inside joke or a fond memory into the conversation, or an expression of love. If you're in the wrong, admit it quickly and change your wording. Most of my apologies go completely ignored by people who get upset with me because I give them so easily and quickly, but they're still there, and I've learned to point them out when I get attacked for not admitting my faults. Then, when they see that I did admit my wrongs, they feel safer having an honest dialogue, too.

Lots of times, we already know what we are doing wrong and when someone else actually tells us, out loud, to our faces (no matter how nicely they do it), we can't stand having to smell our own "turd", so to speak. A natural reflex is to withdraw, or respond with anger. We can go from a loving friend or family member, into a defensive, or offensive opponent within milliseconds. It takes a great deal of courage to admit that they are right. It gets easier the more you do it. Catch yourself from getting defensive and making excuses, or explaining away all your reasons for being your dumb self. Putting up those walls between your loved ones and yourself takes work to pull down later, and most people don't like doing it, because they don't like making themselves more vulnerable than they already are. Fixing the original problem is hard enough, having to fix and repair the walls you put up between each other only compounds the problem.

The gist: Trust each other.

18

Rock Bottom, or... Feeling the Pain of Consequences

Every addict who's ever acknowledged their addiction, and every person who's felt the need to change and improve has hit some sort of "Rock Bottom". That's the point where the pain of the addiction or problem was finally bad enough to push them into getting help. Many people will have several rock-bottoms. And no rock bottom is ever the worst that it can get. It can ALWAYS get worse.

In a classic comic strip from Calvin and Hobbes, Bill Watterson uses Calvin, hanging from a balloon way above the ground, imagining all the things that could make his situation worse, to observe, "That's one of the remarkable things about life. It's never so bad that it can't get worse."

Whether it's losing a loved one, injuring yourself, an intervention held in your honor, a moment of sobriety between binges. Maybe it's a good long look in the mirror and not liking what you see. Rock bottom is where we get our personal motivation to change. We don't like where we are at. We "get busy living, or we get busy dying." It's a moment of choice.

I've known folks who've hit rock bottoms while dumpster-diving for "girly magazines" behind the liquor store, or hiding underneath cars to evade a suspicious police officer outside of a pharmacy. Others found it on their way to jail, or simply realizing they won't get to participate in a religious ceremony because of worthiness issues.

Like the biblical prodigal son who realizes he's eating pig-slop when he could be at his wealthy father's home, it's when we finally breach our own personal morals and snap out of it enough to get help.

Don't count on hitting a rock bottom to get help. If you know you should get it, count that as your rock bottom, because the ACTUAL bottom is way, way, WAY down there, and often involves death or loss of all your personal freedom.

But that's a gruesome thought. Let's go back up a few levels, to things like Rehab and Detoxifying. Those may feel like death, but they are far, far from it.

Recap: It can always get worse.

19

Rehabilitation and Detoxifying Back to Health

Rehab and Detox. I don't mean the detox like health food stores promote. You know, the kind where mysterious accumulated, unnamed, undiscovered toxins get purged from your body's cells by eating some sort of unhealthily extreme diet. There's a place for those, but I'm talking about the real serious kinds.

Often-times, pent-up emotions and unresolved traumatic memories can inhibit constructive thinking and healing. They can manifest themselves as health problems. It's crazy, but there's some truth to it. Emotional problems can, and often do cause physical problems, whether it's because we've turned to unhealthy habits or foods to feel better, or ignored healthy behavior because of depression, anxiety, or the like. Yes, changing our diet and lifestyle can have some improvements and help us take some early steps toward lasting change. More often, though, we need to develop healthy emotional habits and ways of thinking, and the other good habits will follow.

If there is some behavior, substance, habit, thought, or emotion that has held control over your lifestyle, oftentimes the first thing to do is find a way to get away from it long enough that your pre-frontal cortex

—the thinking and reasoning part of your brain— can take full control again.

Rehab facilities specialize in this. Usually you enter some clinic, facility, or hospital where you cannot get access to the thing that's been bothering you. Your body and mind go through some usually painful chemical, hormonal, or emotional withdrawals while the remnants of the drug of choice gets worked out of your system. There's a popular myth that it takes 28 days for this to occur. It goes along with the idea that it takes 30 days to form a new habit. Neither of these is always true, but it's a good start and that idea has helped many, many people. It could take 60, or 90 days. Re-wiring your brain, carving new ruts and pathways for the thoughts to flow like water, can take a long time. It could really take years to get the full effect, but since most people can't spend years in some care facility, they are there to give you a good push-start in the right direction, teach you some new ideas and habits, help you overcome mental and emotional hurdles that held you back before, and then send you on your way back into the wild world of life.

A good program won't just drop you back into life on your own, but give you support, tools, and help, slowly, a little bit at a time, as you can handle it. Any newly sprouted tree will crush under the weight of an unruly kid playing on it, but a large, mature, strong tree can have houses and forts built into its branches easily. What we struggle with early on, will not always be struggles later. We will do well to remember that and let ourselves grow slowly, with help from our friends.

If the thing you need to avoid is a person or group of people that are pushing you into bad stuff, one of the hardest things you'll have to do is get away from that social circle. Sometimes that will mean leaving your home, moving away, and escaping that toxic environment.

As you try to give something up that you've emotionally depended on for a long time, those emotions will surface, in full force, and you will likely be an awful person to be around for a while until you are taught or learn to reign them in, in constructive ways. This is how some people get caught while trying to push along their own recovery by themselves. It's when this stuff surfaces that people think you're STARTING to abuse (if they didn't already know you were). But just like those detox diets I mentioned earlier, abstaining from them for an extended period of time, gives your brain a chance to start kicking all that stuff to the curb.

It's not enough to just go through withdrawals, you and your loved ones (whether it's you or your loved ones, it doesn't matter) will need to learn the right way to act to replace your drug or habit of choice. You'll need to learn the right ways to deal with your negative emotions. That takes outside help. If you could have solved it on your own, you already would have. Don't turn down help from other people. Suck it up, give in, and let them help you.

In truth, once an addiction, or habit has formed, it has created that pathway. It has created a rut in your mind that will likely be there the rest of your life. Though, it may not always. It will always be easy to fall back into old ways. That is where a good support system, a counselor, constructive habits that bring lasting happiness, and principles taught by 12-step style programs work into our lives. If we get our social groups and the environment around us on our side and stay determined even after we mess up, it's hard to lose.

In short: Take out the garbage

20

Social Support: We Get By With a Little Help From Our Friends

It's extremely difficult to get a personal change to stick without the help of our friends and family and coworkers and neighbors. When we form any kind of relationship with someone, we fit into these molds, these slots, or labels that we attach to others and ourselves when we're around others. It helps save brain power to give ourselves these shortcuts to use as we interact with other people, but at times they can be traps. How many of us return home for the holidays and find ourselves getting stuck back into the role of "little brother" or "bossy sister" no matter what's actually changed in our separate lives?

When we make an effort to change ourselves, it is wise to enlist the help of those who care about us. If we're friends with the group that steps outside to have a smoke, we'll lose that relationship to a degree, but we'll either reform that relationship in a healthier way, or form new relationships. Find those we trust and tell them what you're trying to accomplish. Ask for their help. Most of the time they'll be understanding and encouraging, even if they can't be directly helpful. Be honest, humble, and sincere when you approach them. Don't let them put you down, but let them be honest in return.

In Napoleon Hill's famous book, "Think and Grow Rich" he goes on and on about mastermind groups and the role they played in just about every millionaire and inventor that's ever lived. Ed Catmull describes at length in his book "Creativity, Inc." the efforts that the leaders of Pixar took to create honesty and openness among their movie makers. They had a "candor" that let them make so many high-quality and well-loved films. They had often and regular meetings where everyone involved was encouraged to be frank, and honest, in their thoughts regarding a scene or dialogue. It let them fail in safe ways without investing too much time and effort into the failure. Thomas Edison was not some lone inventor tinkering away in a lab, at times he had up to 30 people with him in his lab working out some idea or invention, failing thousands of times along the way as they shared ideas and effort.

Every support group and 12-step group in existence, is a mastermind group of sorts. Family councils fill a similar role. These groups not only give us motivation to improve, but often the ability and answers to our problems we need. If you want to insure that you will find improvement, you will need to find some way to interact with others regarding your improvement. For some this comes easy, for some, it will be the hardest part. Depending on our level of personal motivation and ability, our personality, our past, and our upbringing, we will either be ecstatic to join a group, or terrified of it.

For those of you terrified of such a proposition, I promise it is not as scary as it seems, and it gets easier, quickly. I've watched a great many recovering addicts enter a room, sheepish and timid, sit quietly in a chair, pass when it comes their turn to speak, and slip out quietly at the end. I've personally witnessed dozens and dozens of people tell their stories about how scary it was to walk through the door the first time, then tell how it got easier by the second and third, then how much they enjoyed coming to meetings and looked forward to meetings after just a short while. If they keep coming, it's not long before they find themselves opening up, sharing their fears and emotions, proclaiming their successes, and even giving encouragement to others.

Our brain and our body are designed to ENJOY social interaction, though many of us have either not practiced it, or become "socially obese" and gotten rusty at conversing with others. Some of us were brought up in a home that didn't help us with it, or discouraged social development. With repetition and effort and practice, those pathways in our brain will open up, social hormones and love hormones (serotonin and oxytocin) will begin flowing quickly, and with practice and smoothing out some rough edges, we'll have a nice river of sociality flowing through our mind that we'll enjoy a great deal and will overflow into the rest of our lives.

The hardest part is finding social environments that we can participate in often and regularly. It's hard to find group settings where we are comfortable and where we can improve, not indulge in, our weaknesses. Bars and clubs attract a lot of people. A great deal of our society's sociality exists in those settings, but they have their limits, mostly because most people there tend to be under the influence of some substance or the overwhelming sensory stimulation —the lights and sounds— overwhelm our ability to thinking clearly. We won't be thinking our best in a situation like that. Clothing stores, bars, and clubs sometimes use this to their advantage. Sometimes even churches do, too.

We need to add other types of groups to our mix. Service clubs, community organizations, volunteering, faith-based groups, hobby-groups, and more. Like not putting our eggs in one basket or putting all your retirement savings on one company's stock, having a variety of groups we interact with rounds us out the best. In college, even after I had made up my mind to go into medicine and dentistry, I kept taking Civil Engineering seminar classes just because I did enjoy being around engineers and not just future doctors. I knew a husband and wife in Duncan, BC Canada that each belonged to different chapters of Rotary International just to expand their social circle together. Your circumstances will be different than anyone else's and your wants and desires, strengths, weaknesses, and goals, will help you figure out what you should do.

Our society and species thrives by working together toward a common goal. If, or when, we face a world-shattering apocalypse that destroys western civilization, those preppers who entrench themselves into their underground shelters or disappear into the woods may survive, but those who jump right into rebuilding society by adding their talents to the mix will allow society to bounce right back within a short period, and thus we won't just survive, we will thrive. Like a net or web, when we find attachments to those around us, not as much can destroy us or finish us off. We are stronger together. We should still be careful of those we attach ourselves to. Our friends influence us for better or worse and while flocking birds can be of a different feather, those we flock with tend to shape us into who we will become.

To put it briefly: Choose your friends well.

21

Environmental Ability and Motivation: Change the World Around You

Sometimes, even with the best of friends and support groups, or with the pressures of a personal motivation cracking the whip at us to change, or even with all the knowledge and ability in the world, we will still mess up. We will still fail. Like I mentioned in the first half of this book, sometimes our reptilian brain, our survival brain, our "Go System," dominates over our civilized brain —our "Know system"— and there's just not much we can do to fight a strong temptation when we find ourselves, alone, faced with our weakness. Or perhaps we find ourselves in a crowd of passionate people all going the wrong direction, doing wrong or destructive things.

The answer? Get away from your weakness, or get it away from you. Change your environment. If you want to exercise but can't get yourself to the gym, try getting your own gym equipment at your home, and put it right in the middle of your life where you can't ignore it! Tucking it into the empty corner of the basement may not be the best. What if it's hard to get up and run in the morning? Put your clothes on the floor right in your way so you have to trip over them and change into them without even having to think.

Do you have trouble not hitting the Snooze button for the fifteenth time every morning? Put your alarm clock across the room where you have to get up and walk somewhere to turn it off, preferably by the bedroom or bathroom door where you can just keep walking the same direction away from the bed towards the shower or kitchen. What if you don't care enough to turn it off and are just fine with it buzzing away over there? Get a clock loud enough it might wake up your family or your neighbors! They'll hate you if you leave it going more than a couple beeps and you'll feel that pressure to jump out of bed to turn it off.

Want to quit smoking out the back door? Don't use the back door anymore. Want to quit speeding to work? Set your alarm just a few minutes earlier so you don't have to feel rushed. Or set an alarm 5 minutes before the time you need to be walking out the door. Want to learn about medicine? Go hang out at a doctor's office or hospital and ask to shadow the experts there. Want to stop wasting time on social media on your phone, which is always in your pocket? Uninstall the app from your phone so you can only check on a desktop computer, or get another time management app that limits your Facebook or Instagram usage to some time limit you're okay with. Want to quit drugs? Avoid the friends who help you get them, or who you do drugs with.

Most of us can alter the world around us to a large degree. If you want to make it easy to change and need that extra motivation from the world around you, change your environment. Change your bedroom, change your home, change your commute, change your office or job if needed. Propinquity is the state of being near something. In psychology, it means that you'll do or take care of, or participate in what is near you. If you don't want something around, get rid of it. If you can't get rid of it, move yourself away. Fung Shui is the philosophy that your environment affects your emotions, it may have some good ideas for you.

What if you've changed all that but still can't stay away? Put reminders around you. Use a reminder app on your phone to check on yourself. Put sticky notes where you can't miss them and won't easily ignore them (as they pile up around your computer monitor at work.....). Steven Covey, the author of "The 7 Habits of Highly Successful People" was a big fan of personal planners, or day planners, before smartphones. Write goals or assignments down, give them a due date, stick to that due date, deal with it when it's in front of you. Work with your environment and let your environment work with you.

If you want trick-or-treaters coming to your door on Halloween, so you can show off all your cool decorations, don't live out in the country! Find some suburban neighborhood with a bajillion kids, pick a house on a through-street, not in a cul-de-sac, and use lots of cool lights to get their attention. Do you want to avoid trick-or-treaters at all costs? Go live out in the country, with a really long driveway and barely any markers or signs that you live there. (That'll also save you from having to take care of your lawn and put up Christmas decorations!) If you need help keeping your home clean, make it a habit to invite friends and neighbors over for dinner and a game night so you feel the pressure to clean your living room and guest bathroom once in a while.

Can't stop eating those snacks in your cupboard? Don't buy them! It's hard to eat snacks that aren't there. Or learn to make your favorite snacks. That way you'll make them healthier (hopefully), and the work involved in making them will burn some calories, as well as keep you from making too many of them or doing it too often (again, hopefully).

What if you just can't stop criticizing your kids? What if you just can't leave them alone to make their own mistakes and learn on their own and grow up and mature into a great adult? Give them some alone time. Leave them be for certain periods of time. Find someone else who'll watch them in an appropriate constructive way. Bring someone into your home or your life that can teach you what to do. Read! Learn! Practice!

Do you constantly feel stressed at work or at home? Take a look into the psychology of color and learn how lighting, texture, and color can affect your mood. Read up on organization techniques and time management skills and build those things into your home and office that will reduce that stress you're feeling.

Of course, all these decisions have to be made when you're thinking clearly. When you've had a few moments of sobriety when life is all satisfied and good. Have the courage to put those limits in place, or those changes made, so that when you encounter your addiction, your weakness, your bad habit, your stressor, you've got rules made for yourself and you don't have to make that decision at the moment, it was already made by a more clear-headed version of you.

I'm going to go on a tangent right now, that's a bit related to this. I'm going out on a limb to discuss it. I'm sure many people will take issue with how I describe it and what my suggestions are for this problem. That's fine, I can't please everyone. I hope that whoever dislikes what I'm about to say, takes the time to be thoughtful and disagree not on principle or semantics (how I say it). Ready? Let's go.

Much is said about Modesty. Yup, trigger alert, we're going there! There are plenty of men and women who preach modesty to protect people from sexual assault or to help men (and women) avoid improper thoughts. If we all dressed more conservatively, there would be fewer problems between the genders.

There are also plenty of men and women who feel the opposite, that women (and men) should be able to dress however they like and it's up to the observer to control their thoughts and behaviors. To insinuate that women are mistreated because of their appearance is to promote rape-culture.

Both sides are correct. This is a hard topic to discuss and there's no right answer. There's no place to draw a boundary for what is modest or not. It's no fault of a woman if a man attacks her, however she is dressed. Yes, when people are dressed modestly, it is less of a biological and instinctive trigger, and yes, everyone needs to be responsible for their own thoughts and actions.

There will always be a spectrum of dress standards across cultures and environments. There is huge variation and variety of what people are used to seeing and what they consider immodest. That huge variation is also what gets us into trouble. Not all people are used to seeing so much skin! And like I pointed out early in this book, skin is an instinctive biological trigger for certain behaviors and hormones.

Like every person on this planet, how we dress ourselves is "the right way." Those who show off more than us are "trashy, and those who show off less than us are "prudes." This goes along with our driving habits. Everyone who drives faster than us is a maniac who cares nothing about safety, and everyone who drives slower than us is an idiot. We judge everyone else from our own perspective. It's all we can do. We can try to broaden our perspective, but all of our opinions of this world and universe have one unifying thread: Our self.

Avoiding the triggers that spark the hormonal and instinctive behaviors that all people are vulnerable to is a team effort. We all need to learn to control ourselves and our thoughts. We also all need to learn to be conscious of how we appear and the kind of effect it might have on others. There's no line in the sand that we can draw. There are no boundaries that will fit all places, times, or circumstances. There can't be. Clothing of some form has been present on humanity since we became a species, whenever that was. As soon as humans appeared on the earth, they were wearing something, and it covered the genitals and reproductive organs. Even other hominid species likely wore some kind of clothing before we showed up on this earth. Naturalists like to think that nudity is natural, when it actually isn't. Clothing is the natural state of our species.

Clothing is used for practical reasons like warmth, protection, etc. But it doesn't take much clothing to meet those needs. A skin, pelt, or plain blanket will work for that. Anything beyond that is some sort of expression of who we think we are. We wear subdued clothing to show our humility. We wear extravagant clothing to show our amazing style or our wealth. We wear uniforms to show solidarity and unity (hence the "uni" meaning, "one"). We wear bright colors to influence our mood and the mood of those we'll meet. We wear black to show our grief and help others join us in our sadness. We wear tight clothing to feel sexy and attractive, or to be supportive for lots of movement. We wear shiny clothes to attract attention, or dull colors when we want to avoid it. We wear intimidating clothing to make us feel stronger and to project that image to others. We wear clothes to assert our social rank. Or we wear whatever we feel like and get stuck in whatever social rank we didn't care to choose for ourselves (often not a helpful one).

Clothing has just as much of an effect on ourselves as it does on those who see us. It stirs our thoughts and imagination, and it triggers that in others who see us in it. We act differently when we wear different clothing. I've had fun going to dental conferences dressed casually, or very professionally from one day to the next and noting the difference in how the sales reps and other doctors treat me and respond to me. Waiters and waitresses will treat us different based on our dress and we treat them differently based on theirs.

No one likes to admit that they do. No, of course not. We are all much more civilized than that. Except that we aren't. Even the best of us has to fight that instinct to judge people on their appearance, because it's a useful instinct that has helped us survive for a really long time. The fact that I can even talk about this and you can understand this speaks to the power we have as a species to overcome our instincts, with effort. This conversation, this issue of modesty has been around since before recorded history. It will go on forever.

The truth of the matter is, a person who's looking to aggressively satisfy a sexual impulse will more likely pick someone who's showing more sexuality in their appearance than "the average" person for that culture or that setting. Just like a car thief will more likely pick a car that's easier to steal (1980-1990's Hondas????). But no one wants to be less than averagely attractive, so it's no surprise that people want to look their best. There is no defending a human being who sexually assaults another. That person acts on their own instincts.

So we are caught in a tricky situation. A delicate balancing act. A fine line. A razor's edge.

We need to make our own personal decisions and understand that they are our own. We hold aggressors responsible for their own actions, and we understand that we can help each other, sometimes, by dressing out of respect for ourselves and out of respect for those around us. Sometimes, as a community, a religious faith, a family, or as a society, we settle on and collectively decide what types of clothing we will accept and what types we won't. It's different for each group and there are few hard-fast rules about it, and it should be that way. We just need to be careful when we interact with each other with our different standards, and look at each other as humans first.

Conservatives or religious adherents are often criticized for having strict, limiting rules or commandments to follow, often at the expense of some stimulating behavior or substance. Many, but definitely not all, of these rules and commandments were set in place and followed by those with a clear head and foresight to see that there are a great many substances and habits that appear gentle on the surface but can be very destructive over a long period of time.

Research shows, that whether religious or not, having our own, personal commandments that we live by, is a huge step in overcoming weaknesses and problems. The late David Bowie gave an interview later in his career discussing his alcoholism. He had made a firm decision to never even have one alcoholic drink ever again. Alcohol became such a problem for him, and he learned enough about it and himself, that he decided he could not afford to take a chance to have even just a sip ever again, for fear of losing control over himself and relapsing into old addictions and the habits and problems that come with it. His drinking had caused problems for himself and for many of his loved ones around him. It became important enough to make that rule for himself and for their benefit, too.

How do you think he kept to that rule? Do you think he went and hung out in bars and clubs? Perhaps he had to as part of his career as a pop star. Perhaps that's what led to the problem in the first place. I don't really know, I'm speculating. But I also speculate that he eventually figured out what rules he had to make for himself. He learned the invisible fences he decided he would never cross, to keep himself safe, and with practice he learned how to keep those fences there and not cross them, even when his thinking was less-than-clear. I gained a lot of respect for him after watching that interview. It's a great example of someone who has recovered from a serious addiction and the humility and growth that accompanies it.

In a distilled form: Change Your World

22

Service: Get Outside Yourself With an Out-of-Self Experience

One of the best ways to overcome or deal with some bad habit, addiction, or weakness is to stop thinking about yourself so much. Seriously. You're so vain, you probably think this book is about you.

Well, it is. Just like that old song, this book is about...you, but my goal is to get you to spend as much time thinking about other people as you think about yourself.

Any time you spend effort, energy, materials, or time on someone else's behalf, you are serving them. Sometimes they pay you for it. Sometimes they give you a pizza while you help them move. Sometimes they return the favor. Sometimes they pay the favor forward. Sometimes you get the warm fuzzies for doing it. Sometimes you get a hope for a better future, or the hope for a better afterlife. Service is never, ever a one-way reward system. We always get something for serving other people. So let's get over the idea that we serve because we are selfless. We serve because it's what we are supposed to do. It's what our nature tells us to do, most of the time.

Even work and employment is service. No one wants to do so much service that they run themselves and their family dry. They have to get served in return. How do we keep track of the service we are doing for other people, and for the service that other people do for us? We use money, currency, cash, dinero, dollars, etc.

I'm not gonna turn this into a sermon on the value of capitalism or financial gain, I'm gonna tie it all into overcoming weakness, bear with me while I touch on economics for a moment.

Some service doesn't require much skill or talent and can be done by just about anyone with minimal training and experience. Some service requires a great deal of expertise or risk on the part of the server and thus becomes a more valuable or rare service. All service requires time, thought, and effort. Sometimes we think our service is worth more than what others think it is. Sometimes we think our value is too little. Sometimes we can convince others our services are worth more than they are, sometimes others convince us that we're not worth what we thought.

While it's hard to pinpoint any definite value to a service, it's easy to mislead —intentionally or not— either ourselves or others into giving the wrong value for our time and effort and skill. This is where dishonesty, lying, cheating, starts to hurt ourselves and society. Being dishonest to others begins to trap us in our own world that's different from reality. It creates a disconnect between ourselves and others. If we can convince others of a mistruth, then we can convince ourselves of the same or other mistruths. Going through our life and figuring out what lies we've told ourselves is, like so many other things I've mentioned already, one of the hardest things we'll ever do. Catching ourselves in our own lies is tough!

How often do we pat ourselves on the back when we do some small favor for a person we dislike, like picking up a stray sticky note for the messy guy three cubicles down? Oh, man that slob owes us BIG for that one!

I love the scene in the movie "Fireproof" where the main character, a husband who's facing a divorce, is trying to learn to love his wife again and gets a challenge to do nice things for her. So he gets this great idea of making her coffee in the morning before she goes to work. He is more than pleased-as-punch in himself for being so considerate, you can see the confidence in his eyes that this will save the day. When she rushes out the door without even looking at the coffee, he loses his temper, obviously for putting so much time and effort into such a noble act. He had fallen into such a bad habit of not thinking about her, that even tiny favors for his wife required a huge change in his mind.

Overvaluing our own strengths while undervaluing our own weaknesses is human nature. It is another common problem of all mankind, at least for a lot of us. We call this "Overconfidence" or "Arrogance." There are also plenty of people who undervalue their own strengths while overvaluing others. This looks like "Low Self-Esteem"

Yet, how many times does a friend or neighbor spend hours, or days, helping us fix a car just to enjoy our company when you'd readily pay a mechanic hundreds or thousands of dollars for the same work?

What is the value of some of those unexpected small favors? What about when a smile or a hug is all you needed to get through some stressful moment?

Both the service that we give freely, and the service that we get paid for, is the grease that makes the world turn, that keeps society functioning. We can't live without both. What we are willing to do for free or for money, is entirely personal. Those who have should share, but they should be allowed to do so freely. Or organize their neighbors, community, and society, into an association or form of government where everyone shares similarly. It's up to each person and each group to decide what they are comfortable with for their circumstances and we should be aware of and conscientious of that.

Since we all have at least something, we should all share. Some communities will enforce more sharing, some communities would rather let people share on their own.

Have you watched any of those social experiment videos on Youtube where some aspiring filmmaker or prominent YouTuber pretends to be homeless? The video often go like this: A Young adult, or teenager, pretends to be homeless, begging for money. Eventually, someone stops and shares food or money with them. The filmmaker then stops the generous stranger to interview them. Often times the stranger is in dire straits themselves but understands the value of charity. In return, the disguised filmmaker then pulls out their own wallet and gives a generous gift to the thoughtful donor, at which point they both cry for happiness and their renewed faith in humanity. And we who watch it are happy to see all of this.

An alternative to this is giving a large gift to a homeless person and then secretly following them to find out that they immediately went to their friends to share their windfall with others.

When we live our life with the mindset that the work we do, the effort we give, the energy and time we spend on others is service, it should change how we treat it. Some of the service we do we get reimbursed with money for. Some of the services we do we get reimbursed with love for. Sometimes, we get both, sort of.

It can be hard to feel like your efforts are any kind of "service" when you feel like just another wheel or gear in a giant machine, crunching numbers and shuffling papers from one desk to another. If you change your focus and began to think of yourself as a part of an amazing "machine" or company, or country that does immeasurable good for others by supplying them with reimbursable jobs, providing a service or services for others who need it, you might start to look for more ways to improve your skills, or for better ways to perform your part of the service. It can begin to become engaging.

You will start to use more of your brain to do your job in better ways. You'll begin to develop a passion for something you previously had no interest in.

The ways you'll improve won't always be apparent to many people. Maybe you'll get faster with your math skills, or your typing, or learning the software. Or perhaps you'll learn some managerial skills, some people skills, some empathy that lets you connect with other coworkers and leaders around you. Skills that begin to tie and unite various departments to each other. Skills that improve the culture of your office or workplace by making it a happier place to be, and thus increasing productivity.

A happy place to work is a productive place to work. Any tyrant of a boss can scare people into productive or profitable work, but an inspiring leader or coworker, with a quite a bit more effort and skill, can encourage others to not only work profitably but even more so, of their own free will and excitement for their job.

Find ways to inspire your coworkers, your employees, your direct reports, your clients, your customers and patients, your family members and friends, in ways that fit your relationship to them. Find ways to improve their lives through the work and service you are doing for them. If you can, sneak in some bit of extra, to show that you care about them, because you should, even if just a little bit, for another fellow citizen of this planet earth. Go the extra mile, even and especially, when they are having a hard time with life or are being cranky and mean.

The book "The Four Disciplines of Execution" by the Franklin Covey group, is a management and leadership style that focuses on this from a business perspective. It involves teaching and guiding a team to make those extra-mile goals, outside of the daily whirlwind, that help the workplace become just a bit better, one step at a time. It ends up uniting teams, departments, and even large companies and government branches to a common goal, with everyone pulling together to accomplish great things. It'll also work in families.

So, why should we focus on serving other people? Why this focus on work? My dad always told us as kids, "Work is the best therapy." In a lot of ways he was right. I had a chance to serve an extended service mission away from home. While I was serving other people I felt a bit like a celebrity. I stuck out in a crowd, and most of the people I'd smile and wave at would know what I was about and smile back. The few who weren't happy to see me at least gave me the kind of attention that made for good stories to tell about later. I felt important and needed a lot of the time. When it ended and I came home, back to regular life, it was like waking up from a dream.

My head full of these experiences and memories that changed me forever, and I couldn't really share them with others, at least not in a way that they cared or that could impart the meaning of those memories. It was very much a personal "Hero's Journey" that Joseph Campbell describes and it was pretty much my own, for myself. Despite all the people I helped, it had the greatest impact on me. Then I came home.

I was just a regular guy again. Making plans for school and working on the side. Several other friends I had who served on this mission with me fell into a depression of sorts, some more serious than others. Some of them lost themselves for a time, trying to figure out who they were and what sort of purpose they were supposed to fill that could live up to the excitement they'd left. A few of my friends got stuck in bad habits and addictions they've now spent years recovering from.

My dad, as wise as he is, put me on the work schedule for his small health food grocery store. I was on the work schedule before I even got home. I had just been away from home for a LONG time, and he's making me jump right into working full-time. True, I could've quit working for my dad and struck out on my own, but, it was easy to go with the flow, and I didn't know enough about life yet to feel brave enough to make that choice voluntarily.

I followed his advice and started working and getting ready to go back to school.

It worked great. Having that structure in my life, having to get out of myself to serve customers all day long, having to manage my time so I could still go do fun stuff, was a great answer for me. It was one of the best times of my life, instead of a discouraging let-down. Work was very good therapy.

One recurring theme I saw while I was on this extended service mission was how bad retirement could be for people who didn't have others to serve. Once a person quit their job, once they didn't have to work for their living, the usual routine they fell into was doing whatever they want. After a few years, this easily turns into sitting in the armchair for more and more hours of the day watching TV or reading for fun instead of growth. A few years of that and it becomes hard to get out of the chair. Mentally, physically, and emotionally this is when aging appeared to happen the fastest. With no demands on their time and attention, they atrophied. Mentally, physically, and emotionally, they became sick.

I can't blame anyone for falling into this pattern. It's natural. Work is hard. Serving others, especially when they don't appreciate what you're doing for them, is tough. Like Isaac Newton's First Law says, "An object at rest will tend to stay at rest." Most of us are naturally lazy. It takes effort to live, it takes effort to breathe and eat. It takes effort just to exist, and all the other laws of the universe are trying to wear us down, so it takes extra effort to do better than just existing, it takes considerable effort to thrive and be happy and improve ourselves.

If we come across some energetic 90-year-old who claims that staying active and healthy is easy, don't let them fool you, they have just managed to build up enough momentum while they were younger,like when they were 70 or 80 —even if it was more mental and emotional momentum— to keep them going, physically. Or they learned what I had the chance to observe, that if you don't put that extra effort into your life, you will more likely die a lot sooner, if not physically, at least mentally and emotionally. Or perhaps life put enough pressure on them, they had no other choice but to get up and go back to work and then they realized how valuable and life-saving it is.

My dad's parents were hard workers. They tried retiring while in their 70's, and got bored. So they opened up another health food store and worked it well into their 80's, and even when they physically couldn't work the store, they sold it and kept doing bookkeeping and other tasks for my dad and uncles to keep busy. They were able to live well into their 90's. My mom's grandparents all lived into their 90's. Chances are high, that I'll live a nice, long time as well. 20-30 years is a long time to keep busy after a typical retirement, so I've thought long and hard about what I'll do to keep busy for so long. I don't have too many answers, but I've learned a few things along the way.

My eyesight as a kid got really bad. I noticed during the 1996 Olympics that I couldn't read the text on the family TV. At the age of 15, my parents took me to an eye-doctor. By the time I got my first real eye-exam, I couldn't read the big "E" at the top of the letter chart. That's worse than 20/200 vision. I think my parents were really embarrassed. I don't blame them, I didn't know it was that bad either. I'd been able to get by without too much trouble.

The optometrist I went to as a kid kept a couple issues of National Geographic in his waiting room. One of them had a feature article about longevity and centenarians around the world. I read through that article at least 2-3 times during various visits to see this doctor. Funny enough, I went back to see him as an adult, at least 10 years later. He'd moved his office to another building across the street and he still had the same issue of that National Geographic in his waiting room.

It compared the lifestyles, diets, and habits of centenarians (people older than 100 years) from around the world. Seventh-Day Adventists in Loma Linda, California. Great-grandfatherly patriarchs in Sardinia. Multi-generational families in Okinawa, Japan. They were all interviewed and examined. There was very little overlap of diet and lifestyle between these three groups. They had almost nothing in common regarding their genes, or diet, but the one thing they all had in common was a strong social circle, and a need to keep serving and taking care of others.

One woman, over 100 years old, still drove her own car, and spent a significant amount of time visiting "the elderly" in long-term care facilities, often taking care of people decades younger than her. All the interviewed Centenarians in the article were all active and busy. This agrees with what I said earlier in the book that moving your body also moves your mind, and vice versa. Don't sit still if you can avoid it in any way.

Discussing this and seeing this principle in action has confirmed it's value to a long and happy life. Find ways to keep busy helping other people. If you get paid for it, great! If you don't get paid but don't need the money, great! Find ways to keep learning, and keep getting better. I'll re-quote "The Shawshank Redemption" again, "Get Busy Living, or Get Busy Dying." Helping others is not a guarantee for happiness and health. Age and illness still gets all of us, but it does improve the odds in your favor.

The nugget to get from this: Help other people.

23

The 3 (or 4)-Legged Stool

When you are dealing with more than just a personal flaw or weakness, but a truly strong addiction, or compulsion of some kind, you will need specialized help. This step goes along with what we've been talking about already, just taken to the next level.

It's commonly promoted that the key to long-lasting recovery and self-improvement takes three things:

1. A Professional Counselor, preferably one that specializes in your type of addiction or problem and has similar moral standards as you.
2. Some form of ecclesiastical/spiritual support.
3. A Recovery Program, 12-step program, Religion, or program that promotes similar principles.

Like the number of legs needed to keep a stool upright, three legs are the minimum needed to stay standing upright. Hopefully you get the picture of an old-timey stool, but, besides the odd dairy farmer who still milks by hand, who else sits on three-legged stools? We'll go with four legs. Just about everything we sit on has four legs nowadays. Gone are the times when utilitarianism and thrift prevented us from a fourth support. Here it is:

4. You will always need the support of family and friends.

As a caveat before I go in-depth into this in a bit: if your family and friends are the cause and source of your bad behavior and addiction, then you have to weigh the need for possibly severing yourself from them and finding and creating new relationships with others who can and will help you. There are few things that could be more painful than severing a long and cherished relationship, especially with a spouse, or family. But sometimes, that is what is required and can, at times, be better for both sides. I don't know what grounds or rubric should be used to justify that decision. Infidelity? Abuse? Neglect? I've known relationships to survive and even THRIVE through moments of these things, while most collapse under them. Deciding what you will do will be completely individual. Sorry. That's just how it's gotta be. You'll figure it out. I'm sure you will.

Back to the stool.

24

Professional Counseling: Because Two Heads Are Better Than One

Professional Counseling involves going to someone who has training involving how people think and feel, who can help you sort out the mess of thoughts in your brain. This is supposed to be an expert in helping you understand yourself and other people. They mostly listen while allowing you a chance to say whatever is on your mind about the topic you're worried about, and then they ask you probing or open-ended questions to get you to explore your own thoughts and memories further. Think of them like a guide on the safari of your own mind. Their job isn't to solve your problems, but to help you find ways to solve your own.

Counseling can be a wonderful thing for many people and often will be a key in overcoming past difficulties, or present addictions and behaviors. A private counselor can help you in ways that no one else may be able to, and you'll be able to discuss things with them that you might not feel comfortable discussing in any other setting. But understand, many people have difficulty finding a counselor who does help them. It's very individual. Personally, I recommend it, but it may take you a couple tries to find one that works for you the way you need it.

Don't feel like you have to stick with a particular counselor unless medical or other considerations warrant it. Psychiatrists have medical training and can prescribe prescriptions that they feel might help you. They have a tendency of relying on such methods more often than other types of counselors, which can be a great thing for many of us. Psychologists are similar, but cannot prescribe medications. Both can prescribe various psychological tests and interpret the results. These tests can help you and them understand your personality and emotions and come up with a plan of action catered to you.

There are various addiction counselors, therapists, and marriage and family therapists with other licenses. They are mainly there to help you work through your thoughts, feelings, and memories, give you exercises or suggestions to help you improve yourself, or help you explore your thoughts and feelings. Some have experience and certification in hypnosis. Hypnosis is merely a state of deep relaxation and trust, where you are more open to, but not bound by the suggestions of another person. You will always be in control of yourself, but if done properly, you will not be inhibited or afraid from doing what they suggest. By doing so, they can help you overcome some of your mental blocks, and thus make decisions you wouldn't normally make due to fear or indecision, or various irrational excuses your fully-alert mind uses to protect itself from change.

The first deaf-NBA player, Lance Allred, gave a fascinating TED talk about his life growing up in a polygamist community. He makes the profound statement, "Most people would rather choose a Familiar Hell, than an Unfamiliar Heaven." That's very true for many people. A good counselor will help you figure out what is holding you down, and help you feel safe enough to explore what could lift you to another level.

Counselors and psychologists come from all backgrounds, cultures, belief systems, and religions. While medical training and licensure offers some uniformity, there are big differences between the counselors and therapists you can find out there. Be conscious of that as you meet them and decide on one over another. Some are taking on the title of "Life Coach" nowadays, though not all "Life Coaches" have had adequate training.

Having someone else who can help you through your own thoughts is like having a flashlight out at the end of your arm that can show you where your thinking is contributing to your problems. Having another person -a different, unrelated, and unattached person- give you a unique perspective on your life is incredibly valuable. It's necessary for many of us.

Here's the rundown: Use experts.

25

Ecclessiastical and Spiritual Support: Setting Your Sights on the Stars

Spirituality and religion are touchy subjects for many people, but it is a nearly universal concept inherent in humanity. I'm not here to endorse one version of faith over another. I'm most familiar with Christianity, but I understand that everyone who reads this is going to come from a different perspective, set of experiences, and general feel for life, spirituality, and religion than me. That's a good thing. Everyone has had different experiences with their life and the universe. Take your own experiences and add to it.

Many people benefit from having a spiritual guide of some sort in their life. Whether it's a personal study of religious texts and scriptures, an organized religion of some form, or a loosely based belief system shared by many others. Most people on the planet believe in some form of higher power that exists outside the constraints of our physical universe as we understand it, or that a higher intelligence is possible, even if we don't understand it, or haven't yet seen it. Many people believe that that power or entity can guide them, communicate with them on some level, and influence the world around them to some degree.

For most people, this belief gives them hope, a desire for improvement, an example of how to live life, or instructions on how and why to love others and the world around them, whatever its condition. At the very least, some of us believe that there is a generalized concept of "good" that we should work toward. Some call this "God," "Heaven," or something similar. Some don't know what to call it, like Agnostics. Some call it "Intellect," or "Progress," or "Intelligence" like many of my Atheist friends.

Whatever you call it, there is a sense of peace and security that comes from it. Often we can receive a sense of escape from regret and past mistakes, or help dealing with and coping with regrets and past mistakes. This is vital for anyone looking for long-term recovery. We all need some way to escape from under our guilt without simply ignoring it. We have to have a healthy scapegoat of some kind, or else guilt, despair, cynicism, hopelessness, and pessimism will rule over us and we'll fall back into our old escapes, fantasies, and addictions, downward and downward. A scapegoat isn't something we blame for our mistakes, it is an object, entity, or idea that we can pass our mistakes onto, knowingly, so that we can let go of them and move on and progress. Of course we need to do our best to fix our mistakes and make amends for the damage we've done. There just comes a point where we need to move on and put it behind us.

I'm not so concerned with what religion anyone is right now at this moment. Whatever religion or belief system you belong to and prescribe to I only recommend you develop a personal relationship with It. Attach yourself to anything, from whatever source, that healthily gets you closer to God, the Universe, Nature, Eternal Mother, Nirvana, Enlightenment, Heaven, Home, or Inner Peace. To me, for this book, these things are all synonymous and equal. Whatever helps you and those around you get closer to these things, is Good. I know that there's deep doctrinal and ceremonial differences, but for the most part, here on Earth, the end of it results in Hope for the future. If we live up to our beliefs, our various behaviors start to become very similar and we find we have a great deal in common.

Beyond our own personal relationship with our own spirituality, religious leaders have often had a fair bit of experience helping people with complex deep personal and spiritual matters. Few are formally trained in psychology or counseling, but experience and their religious training can sometimes make up some of the difference. That said, religious leaders should not replace professional counselors, but should supplement them.

Religious leaders can help you find areas where you may be deficient in practicing your beliefs, or areas where you are strong and can take encouragement. Religious leaders should not encourage you to hurt yourself or others directly or intentionally (unless it is incidental to your own healing like getting away from a toxic or dangerous relationship, or how changing your environment can cause pain and discomfort to yourself), but should be able to help you find ways to help others. Especially, those who are different from you and those who may have hurt you in the past. Just about every religion out there believes in and encourages forgiveness and the power of forgiveness and your ecclesiastical leaders will and should help you learn to forgive yourself and others. More on that later.

Any ecclesiastical leader who encourages you to take property from others, hurt them physically, or be verbally or emotionally abusive and harmful, is operating from a place of defensiveness, or fear. Be careful to figure out which is which. I've already described how dangerous fear can be. Even defensiveness can be destructive when used inappropriately. Religions or philosophies of any kind that we use as guides for life need to operate from a place of Love and Hope, for both the immediate future and the long-term survival of humanity and the individual.

In essence: Aim High

26

Recovery Programs Are Steps to Better Things

Most recovery programs I am aware of incorporate and involve the 12-Step method to some degree or another. Even a lot of religions have been using many of the same principles since before the 12 steps were formalized by Alcoholics Anonymous. The first time I read through them I thought to myself, "This is what my Sunday School teachers have always been telling me!" Whether it's through a formal "Anonymous" program (Alcoholics Anonymous, Sexaholics Anonymous, etc), through another support group system (Weight Watchers), whether it's a paid-for system, or a free community service, the principles that are embodied in the 12 step program are fairly universal. Does a program have to have 12 actual steps to be effective? No. Definitely not, but it should have some resemblance to the ideas and actions that are fulfilled in the 12-step programs. The twelve steps have a great deal in common with many of the religious teachings on self-improvement. The original 12 steps themselves were very religious in nature, and religious self-improvement has a lot in common with the ideas described in the 12-step programs.

That said, for the atheists and agnostics reading this, I'm not trying to neglect you. Just understand, culturally, where these are coming from. I'll translate the religious stuff as best as possible.

There are a LOT of support groups and "Anonymous" programs out there. Here's a list of what a quick internet search will bring up: Alcoholics, Narcotics, Crystal Meth, Marijuana, Cocaine, Nicotine, Pills, Food Addicts, Overeaters, Sexual Compulsiveness, Sex and Love, Sex Addicts, Sexaholics, Emotions, Clutterers, Hoarders, Debtors, Workaholics, and Co-Dependents. These are just groups that have "anonymous" in the name. There are even more programs and centers loosely based on the AA program all over, and dealing with all kinds of problems and issues.

While it would be fun to rehash and recap and basically summarize the 12 steps in detail, I'll leave anyone interested to go look them up yourselves. In very short, you'll need to be honest about your problem, trust others to help you with it, find someone you can trust to tell it to, find ways to let go of the anger in your life, stay the course, and find ways to help other people get better.

The whole idea of the 12 steps is a focus on personal improvement and once you've started on the path of improving yourself and fixing what faults and mistakes you can, go and help others get better, not by forcing, scaring, or intimidating them to get better, but by helping them in their own way to get better, and sharing your successes with others.

To rehash this: Use a system.

27

The Easiest Person to Lie to

You need to be Honest with yourself. This takes a great deal of personal responsibility. It takes a brutal honesty with yourself. Many authors, speakers, and life coaches talk about the "Stories we Tell Ourselves." Those quicker-than-lightning reasons for behaving poorly or for getting mad and upset at others. They usually happen so fast and so automatically that you don't even realize they are happening. But they are. When you can learn to catch those stories before you act on them, you can begin to steer them and correct them. You can learn to control your own mind and body, and thus you learn to control your world. And when you can control your own world, you'll start to have tremendous influence on the world of others and the universe around you. Not through mystical, magical powers, but more often, through the sum of all your actions and thoughts added and compounded together, like a snowball that just keeps growing.

Here's an example of a story we tell ourselves. Or, of a story I've told myself more times than I should admit.

I'm driving down the freeway (or "Interstate", whichever you prefer), and a bright red muscle car comes up behind me, and gets so close he fills my entire rear-view mirror with himself. Riding my tail. Being real obnoxious. I tell myself, "Ugh, so rude. He's freaking me out, but he's not gonna get the best of me, I'll show him." So I actually let off the gas and slow down, maybe I'll even step on my brakes and slow down further. Just to stick it to him and show him who's boss. That's right. I AM. That poor, impatient sucker is just gonna have to deal with it.

What happens next? He gets even closer. Maybe he honks at me, or flashes his lights at me. Maybe he swerves around me and flips me the middle finger as he drives past.

I tell myself, "Ha, that jerk GOT what he deserved, I hope some policeman spots his stupidity and gives him a ticket for speeding and driving recklessly!"

How many of you have had some experience like that? Just about everyone who's ever driven a car. Don't lie, you've all been there. What's the story I told myself? Something like this, "This driver is a totally entitled jerk who's inconsiderate of others, loves to show off in order to compensate for other things in his life, and drives recklessly."

What if we could tell ourselves a story like this instead, as if you were the other driver? Perhaps you've been the other driver at times: "I like my car, it fits my personality. It also matches the kind of people I work with and live with. I'm in a bit of a hurry a lot. Sometimes it's because I was helping my wife take care of sick kids before I left for work and now I'm running late. Maybe I'm in a hurry because I got an unexpected call from work to be somewhere at a certain time at the last minute. Sometimes, I just drive kinda fast because it's fun and I never go more than 5 or 10 mph over the speed limit, nothing dangerous, just because I really enjoy the fun of driving.

"Well, here's some driver going exactly the speed limit, (even a little under the speed limit sometimes) in the Passing Lane, blocking traffic behind them. I'm just gonna get a little close, kind of nudge them awake a bit, let them know that others need to get around them. WHAT'S HE DOING?

Brake-checking me? Don't they know how dangerous and dumb that is? Crazy! I better get around this guy as soon as possible! They don't know how to drive! How do I let them know that what they did is totally inappropriate? I don't know, but I'm gonna let him know that I definitely didn't like it!"

And we flip em the "bird".

All this happens in seconds.

What if instead we were able to catch ourselves telling ourselves these stories. What if, instead of thinking that the "other" person is acting stupidly, we manage to take a split second longer and check all the circumstances and realize, "Oh, I'm going kinda slow in the passing lane, I'll just put on my blinker, move over, and let him get on his way. Oh look at that, he even waved thanks to me! What a nice guy!"

And in return, "Oh, here's some 'grandpa' in the fast lane, I'm just gonna get in his space just a little to let him know I'm here and back right off and give him a minute to feel safe enough to move out of the way. Hey, he did! What a good guy!" And we wave him a thanks as we move on.

Imagine how healthier we'd be if we didn't get our heart rates going, our teeth clenched, and our stress acting up! Imagine how much better our day goes, just by catching the automatic story, and taking control of our mind.

We didn't have to make up any story about the rude driver being late to the hospital, or make up some imaginary tale about how the other driver is a completely sociopathic narcissist. We take a moment to remember the fact that these pretty normal people we're encountering in a pretty normal setting are just in a different mode of life than we are at that moment.

We tell ourselves these same stories to justify acting stupidly to ourselves. "Yeah, I can afford to eat a 3rd piece of cheesecake." "It's just one more joint with friends, I'll be fine." "I'm only a little buzzed, I can still drive." "I'll go to bed in just a few more minutes..." These one-lined stories whip through our brain so ridiculously fast, almost no one recognizes them until some outsider points them out to you, or gives you the idea that they even exist so you can start to watch for them inside yourself.

Being totally honest with yourself is difficult, and takes a lot of practice, and failure, and more practice. Sometimes we even have to get detailed enough to write down our "stories" and self-lies just so we can look at them like an outsider. Sometimes we need to get outsiders to look at our really personal and private "stories" and help us see them for what they are. Friends and family that encourage us to keep doing stupid things and persist in our stories don't make for very responsible or helpful family members or friends.

Validation and vindication is addictive in itself. We can't wait to tell our sisters, coworkers, or close friends about that awful person who snubbed us at the party, or the awful rude cashier we had to deal with at the grocery store, or the incompetent scum of a human who cut in front of us at the Costco Gas Station.

Find others who want to help you get better, at least, in the ways that you need help with.

Once you've learned how to be honest about your problems, you can start attacking them. You can realize that you need others' help to deal with them. You'll realize that others can help you and want to.

Once you've got a grip on your problem, or problems (because who has only ONE anyway?), it'll take serious effort to keep working on them. It might be 3 steps forward and 2 steps back. Or some days it'll be 1 forward and 5 steps backward. Just keep taking steps forward whenever you can. The direction you're trying to go is more important than the progress you are actually making in it. Understand that life is going to keep happening.

There's lots of good stuff in this book to re-read and review, and lots of other books I've shared ideas from that are worth reading, too. Write or draw or record your thoughts and feelings somehow. Create. Build. Craft. As you do, take stock on what comes out of your head and why. Take the time to figure yourself out, while you do this.

Whatever you do, find someway to get what's in your head, out of your head. Talk about it out loud and record yourself. Or tell another person. Write it down, in great detail. Draw it. Sculpt it. Sing it. Get it out of your head and in some form or shape that you can look back at it. All of it. Good, bad, and ugly. Put some physical form to it that you can analyze, and grapple with, and hold. And be careful with it. Not many people are ready, equipped, or able to deal with your worst parts, distilled down into such a concentrated form. Don't pull your worst out, just to spill it out onto everyone else. Find someone or a group who can help you properly deal with and dispose of that kind of waste.

Just like putting a name to our fears, we need to put a description and shape to our thoughts. They remain nebulous and wispy while they stay floating around in our heads. We can't do very much with our thoughts until we begin to pull them out and put them down on paper, or something. Only when we can look at ourselves like an outsider looks at us, can we see ourselves for who we are and then do something about it.

When you've got a problem that has persisted in your life for as long as you can remember, you HAVE to find some way to pull it out and look at it and figure it out. Like pulling out a big cancerous blob of black slime that has tangled itself in between the nerve cells of your skull. Or the roots of an obnoxious weed that you don't want among your flowers or garden. You can't get rid of it if you leave it there. You may not get all of it at once, but as you pull each new sprouted plant out, you will weaken it, and weaken it, further and further. It may crop back up when you least expect it, years in the future. That's okay. You'll know what it looks like and you'll have had practice pulling it out, or you'll at least have been able to carefully quarantine it to a section of your yard where it does serve some useful purpose. But you do need to acknowledge it, put boundaries on it, and recognize it for what it is.

This is a hard and painful step. We don't want to see our ugliness. We don't want to pull the muck and slime out of our brain and actually have to look at it. It's like finding the moldy spaghetti in the container in your fridge that's been there for too many weeks. Forcing yourself to look at it and deal with it is CRITICAL for growing and getting better. If you want to help other people in this world, you need to clean out the refrigerator of your mind. If you want other people to help you, you need to be safe enough for them to get close to you to help. Clean out your garbage, air out the stinky bedrooms. This is a step you'll have to do yourself, but it's the most important step for being able to connect with other people.

In fact, it's in the building and writing and creating of relationships that you will figure out who you WANT to be. Building relationships requires correcting them and fixing them when you break them or strain them. And learn to be accepting and encouraging of others when they break them or strain them. As we form connections with other people, we will strain those connections at times. Acknowledge that, apologize for it, and fix it whenever you can. What you'll find is that your connections with other people will grow stronger after a few of these repairs, and you'll learn the limits of each relationship and you'll both learn in what ways you can depend on each other, safely and reliably.

When you're wrong, admit it quickly and easily. The faster you do it, the easier it is. Make up for your mistake, fix it, if you can. Apologize. Learning to say, "I'm Sorry", or even, "My bad" is just as, or perhaps more important than saying, "I Love You." It doesn't even have to be a full apology. A simple, "Excuse me, what I meant was..." can have the same effect. A quick, "Oh, you're right, let me rephrase that..." can go a long way.

There's a popular idea that "Love means never having to say 'I'm Sorry'". I'd like to bury that idea. It's an awful idea. Love is learning to say "I'm Sorry." Equally important is learning to say, "I forgive you" or even, "That's alright, I still love you." We will all make mistakes and we all need to take the time to correct them. And when others tell us that they are sorry, accept it and forgive them and move on. Don't hang onto anger. Let sadness and regret go.

Sometimes we get hung up on wanting apologies from others. We want apologies from corporations or governments, we want apologies from leaders for what their followers did, we want apologies from our spouses or parents or neighbors. You may or may not get those apologies. Don't suspend your life, don't hang your emotions, or hold your breath waiting for those apologies. Whether you get the apology depends almost entirely on the other person and whether they can recognize what they did. We might help them recognize that in a safe or loving way that doesn't cause them to drag their heels and hold their position, but trying to coerce or force an apology out of someone else only stirs up our own anger and resentment and causes them to feel anger and resentment. Do what you can and move on. Let it go as soon as possible. If there is a pattern or repetition to their mistakes, study it out yourself and try different ways to help them recognize their behavior and its effects on you and those around you.

Have trust that even if you can't help the other person see the damage they are causing, they will likely see it someday in the future, maybe with your help, maybe with someone else's. Don't put your life on hold waiting for someone else to make the right choice.

Focus on your own right choices. Recognize your own depression, fatigue, stress, boredom, loneliness, or anger. Try to figure out what triggered it. Make lists of the triggers you encountered during the day and address them.

Some people feel like taking this kind of daily inventory is tedious, time-consuming, or not worth the effort, but it's a lot like putting gas in your tank or oil in your car. If you don't take the time to do it, dealing with the aftermath of NOT doing it is much worse. This is one of the things you'll learn to do to keep yourself clean and healthy. You can learn that it makes your life not just clean, but a lot better.

Many people recommend keeping a diary or journal. Keeping a journal is a daily personal inventory. Learn to focus on your own improvement, not on trying to fix everyone else. As you get better, you will find ways to gently help others get better, or they will see your life improve and they will want to know how you did it. Your light will shine to them. Give the credit where it's due: on the world, or experts, or God, or inspiration, or wherever you feel like you got it from. As you give people credit for the help they've given you, they will feel proud of themselves, and love you more for your acknowledgement of it. You will form new friendships, new family members, and new connections with those around you.

You can only take credit for so much of your own improvement, just as you can only take credit for a portion of your faults. Make sure you acknowledge what is what, and learn to tell yourself the truth.

The quick version: Tell the Truth

28

Filling Our Lives With Good Things

When we give up one bad habit or addiction, it's extremely easy to fall into a new one. Our life has a new void that used to be filled with some indulgence. We have to replace it. It is like putting a layer of mulch over the top of the soil. In gardens, this'll do two things: Keeps water and nutrients in the soil from washing away, and prevent sunlight from getting to the weeds and seeds we don't want. It's not impervious or permanent, but it helps.

Where I live, sage brush and thistles will grow on any spot of ground that is left alone. Where my in-laws live, great big trees will grow anywhere they are allowed to. The weeds we deal with in life are very dependent on where we live and what our environment is like. We need to learn to protect ourselves from future weeds, or future problems. Just yesterday I heard someone telling me about their frustration keeping dandelions from growing in their lawn. A neighbor wisely let them know that it's not about fighting against the weeds, it's about strengthening the lawn.

Let's keep this weed and gardening analogy going for a bit. This is just a way to think about, picture, and imagine our lives.

It'd be easy to get into a discussion on what IS a weed and whether we should do anything about them. Thoreau and others will claim that weeds are entirely social constructs. And like weeds, our lives should grow how nature inclines them to grow. There's some truth to this, but it's not complete. Most plants that we call "weeds" grow only where we've disturbed the environment, and since it is impossible to live on this earth without disturbing our environment, most weeds are only here because we are here (did you know that the iconic western Tumbleweed isn't native to North America?? It came from Russia in the 17-1800's... It's an invasive species. Oops). Weeds, both the plant type, and the emotional types, have evolved along with us, and thus are invasive species wherever we go and wherever we grow. Some weeds can be useful, some are problematic.

Once we recognize that we are the ultimate masters of our own life, that much of it is in our control and only goes and grows as we either consciously decide, or as we fail to decide, then we need to recognize that every aspect of our personality is some sort of construct of our DNA, our environment, and our decisions. No single one of those is in full control. We cannot control our DNA, but we can learn about what genetic tendencies we've inherited and make accommodations for them. Short women tend to wear high heels, while many taller women elect not to. Those prone to skin cancer should use various methods to block the sun, or avoid too much of it. Those with a more natural hand-eye coordination will often be more athletic from an earlier age. Those with excellent memories and analytic intellects may have to try harder to pick up on social cues. Our DNA has a strong influence on us, but doesn't get the final say on how we act.

Our environment is out of our control, to a degree. We can choose to live (at least in the USA) where hurricanes or earthquakes are common. We can choose to build our homes to withstand hurricanes and earthquakes. We can choose to wear the clothing suited to our environment, and eat the foods most easily grown in our environment. When a drought affects a society far away, we can choose to share our resources with them. When a flood washes away our belongings we can choose to reach out for help and aid, and rebuild what was lost, if possible.

Our decisions are our own and we are responsible for them, or the lack of them. I do not mean to say that we are responsible for resolving every problem that our DNA and the environment have thrown at us, but that we are responsible for working towards improving our situation wherever we are. If we are totally and completely dependent on the care of someone else due to health or disability, we can find ways to make their task easier, by being friendly, grateful, conversational, appreciative, and caring.

In the end, we get to choose what types of plants we will or won't let grow in our garden. Which habits or lifestyles are the weeds? What do they provide for us? What do they do that is beneficial or harmful to our neighbors? That is up to us to decide, and we will have to live with the results of those decisions, and keep working to change and alter our lives as we can or as we see appropriate. And we have to let others do the same. We are as in control of our lives, as we are knowledgeable about who we are. The less we know about ourselves, the less control we have over our lives. When you get rid of one thing in your life, you will have to replace it, or you'll be dealing with whatever natural habits or weeds can fill that space.

Bill Wilson, the man who began Alcoholics Anonymous was able to achieve lasting sobriety from Alcohol, but his forays into LSD experimentation (a brand new, recently-discovered drug at the time), his affairs with other women, as well as his addiction to smoking despite suffering from —and eventually dying from— emphysema, hint at his inability to subdue other addictions and addictive behaviors. I don't mean to knock the contributions of Bill W, nor demean him as a person, but hold him up as an example that we all need help even after we achieve sobriety. He helped millions of people, personally, and through A.A. and left a lasting legacy. But it is also a warning that we cannot let our guard down for the other dangers that will want to crop up into our lives once we have removed the original cancer.

We must fill our lives with good things. We must fill it with constructive behaviors as best as we can. Life will never be perfect, nor easy, for anyone. I've seen both sides of society to know that some of the poorest people out there are the happiest because of their relationships and work ethic, while some of the wealthiest are the most miserable because of their broken families and selfish pastimes. Of course there are the poor among us with devastated families and there are wealthy people with amazing lives, but we all have challenges. It could be health, it could be emotional, it could be financial. If you don't have huge, life-shattering problems, just wait, you will (or you'll be lonely, which can be life-shatteringly unhealthy whether you see it in yourself or not).

The only way to overcome these problems and rise above them, is to take the time to carefully think, ponder, pray, meditate, study, and learn what you should do with your life. This overlaps with taking daily inventory. Taking the quiet moments to address your life and where it should go next.

For a good part of his life, Bill W wasn't a believer in God and he was determined to stay that way. But while in a hospital on account of alcohol, threatened with an impending death if he didn't do something, he reportedly begged the universe for an answer or help that he hadn't been able to receive before. A close friend and former alcoholic, "Dr. Bob," recommended he turn to God and Christianity. Bill decided he was desperate enough to try anything and began to pray. He had some sort of profound experience while doing so that convinced him that there was a God and he actively pursued that God from then on.

Anyone who pursues a relationship with the Divine should tell you that the pursuit doesn't qualify you for perfection, nor does it make you perfect, but that the quest for perfection keeps your life going in an upward direction that at least keeps you from falling back into old problems, or gives you the hope and motivation to pull yourself out of them again and again and again.

Many people have felt or experienced God, nature, the Universe, their subconscious, or *something*that talks to them in quiet moments of reflection. It gives flashes of inspiration, ideas, memories, that prompt them upward. Whatever the source of that inspiration is, or whatever we think it might be, look for that inspiration. Take the time to quietly and soberly seek it out and find out what we should do to help others, to help ourselves, and to continue growing and serving, instead of indulging. It will be different for each of us. Very often, it will involve helping others and being involved with people. Do you sense a theme here?

Fill your life with good. With helping others, and taking care of yourself. Exercise, eat healthy, learn new skills. Breathe. Control your breathing and slow it down. If you can control your breathing, you have more control than most people. Form relationships and do what you need to to grow and add to those relationships. Plant the flowers, fruit trees, vegetables and grains in your life that will allow you to grow and feed your neighbors. Expect weeds and pests to keep trying to infiltrate, and stay on top of them. They are much easier to deal with if removed while small and weak.

Synopsis: Plant good seeds.

29

Spirituality and the Golden Rule

Spirituality is a sensitive subject for some people. Some people have found spirituality on their own, some found it through their religious practices. Some find it in the wilds of nature, some find it in magnificent churches. Some find it in the quiet of their personal rooms. Some find it while praying, some while reading, some while volunteering. Wherever YOU find it, hold onto it. Listen to it. Follow it. And do what you can to keep your own ego, your own selfishness out of the way so it can continue to flow into your life. This will take incredible practice and you will likely struggle keeping it that way your whole life. You will still fall. You will still mess up. Maybe not in your original addictions and problems, but perhaps in new ways you haven't considered or imagined yet. Living life is messy and complicated. No one is very good at it, and those who appear to be good at it will usually tell you that they're just good at hiding it, or they've had a lot of difficult experiences that gave them the necessary practice to get where they are.

The key result of all true spiritual experiences is a new or expanded love for yourself and a love for those around you. Everyone around you: your politically backwards neighbors, your dysfunctional family, the citizens of far away countries who stomp on your flag, everyone.

Our brain is designed to care about our own body and acknowledge how our body and its actions affect those around us. Oftentimes, a trauma, an experience, a substance, or something else can hinder or hurt our brain's ability to observe what is going on inside, or perhaps outside. We can become blind to our own needs, or blind to the needs of others. The worst part is we can't even see our own blindness. Empathy and a clear head helps to restore our mental sight to both sides.

What causes some parents to abandon their young children, or teenagers or adults to participate in risky behavior that seriously endangers their life or the life of their family? It can often be traced to some sort of trauma, substance abuse, or addiction. Some environmental trauma (whether through a substance, or set of emotional experiences) infiltrates our minds and kills either our own survival instinct, or our biological drive to care for our families. Sometimes it's just plain bad information or education. Sometimes, it's just plain immaturity and inexperience.

Some people think that happiness is the purpose of life, but the continuation of life is the purpose of life, and usually we find the greatest happiness or fulfillment in helping life continue. I'll repeat that.

Happiness is NOT the purpose of life.

Continuing and Enriching Life is the Purpose of Life. This is Love.

Love is helping life continue. Our greatest Meaning comes from promoting Life, especially Human Life.

As you help life perpetuate itself, you will find fulfillment and, at times, Happiness. In fact, in helping life continue, at a better quality than it is now —especially human life— you'll find more happiness and satisfaction than in any other way. Much is said about "Meaning" in life. People find meaning in their work, their families, their pastimes, their activism, their charity work, or the legacies they leave behind. All of these are unified by the goal of helping humanity continue and prosper. If we are looking around the world to find "Meaning," to fulfill us, we will never find it. Meaning comes to us as we fulfill the purpose of life for ourselves and others.

C.S. Lewis wrote in a letter to Mrs. Johnson in 1955, *"I think I can understand that feeling about* ***a housewife's work*** *being like that of Sisyphus (who was the stone rolling gentleman). But it* ***is surely, in reality, the most important work in the world****. What do ships, railways, mines, cars, government etc exist for except that people may be fed, warmed, and safe in their own homes? As Dr Johnson said, 'To be happy at home is the end of all human endeavour'. (1st to be happy, to prepare for being happy in our own real Home hereafter: 2nd, in the meantime, to be happy in our houses.) We wage war in order to have peace, we work in order to have leisure, we produce food in order to eat it. So* ***your job is the one for which all others exist.****"*

In essence, perpetuating the human family is the reason we work at all, in whatever way we feel like, or have the opportunity to do.

Too many people wander all over their life looking for something that strikes their passion. People want a job they can be passionate about. They want a spouse that stirs their passions. They want excitement. They want all their senses and hormones fulfilled by the world around them.

Mike Rowe gives a great snippet on Youtube about how following your passion is a bad idea. Cal Newport in his book "So Good They Can't Ignore You" describes how, against our usual ideas, passion comes with skill and experience, not the other way around. In other words, we want passion fed to us, but instead we have to go looking to become passionate about what we're doing. This seems to go against what I mentioned earlier about Shawn Achor's research on how Happiness leads to Success. There's a subtle difference between Passion and Happiness. They are different things, though they can stir up each other.

We need to find happiness inside of ourselves. We need to be satisfied with who we are and grateful for what we have. Then, we go looking for ways to help others, including our children or future children. As we get better and better at what we do in our own specialized field, we become passionate about it as we see the value we've added and the benefits we've created. Whether it's in growing food, crunching numbers, creating art, or pushing the limits of science, technology, and entertainment.

Remember, Passions change, depending on our environment, opportunities, and life-circumstances. There's no such thing as a Soul-mate, though there are people that we get along with better than others that are as committed to us as we are to them. There is no such thing as a true-life's calling, though there may be unique opportunities presented to us that fit our circumstances very well. Passion is created over time. Many studies have shown that people are their most productive in their careers while in their 50's and 60's. Beyond that, many people don't NEED to work anymore so they eventually get tired of working, looking to fulfill their own desires.

What keeps some of the very wealthy and very old working past their need to work for financial security? Is it simply their greed for more? When you observe them, it doesn't look like it. They find their fulfillment in helping others, in whatever way suits them, in charity, in work, in family, or in church. This is what keeps them going.

For people who are no longer having or raising children, helping your grandchildren, or other children in your community has a huge impact on yourself and the families around you. Like I've stated earlier, those who stay socially involved in their communities live longer and happier lives. It can be hard staying up-to-date and relevant enough to stay involved. It will remain more and more "work" to keep pace with life and keep up as best as you can, but it's worth it, for everyone you'll meet, and especially yourself. Several studies have shown that children who grew up around their grandparents were more likely to live longer and have more children. Our species is built on multi-generational families, with support from bigger communities. We do better, together.

I saw a Meme floating around the interwebs not long ago that said something like, "If you raise your children, you can spoil your grandchildren. If you spoil your children, you'll have to raise your grandchildren." This isn't a universal truth, of course, there are lots of factors that cause troubles for families, but there is a lot of truth to this.

Do you find it hard to relate to younger people? Is their music nothing but noise, and their clothing completely awful and weird, and their movies just pure garbage? That just means you're old. Give it 60 years and the best of today's styles, music, and entertainment will filter off and distill down to the very best that we remember and keep around. Then all we'll see of what life was like when we were younger is the purified classic stuff that was actually good. There was junk music and fashion around when Mozart and Beethoven were composing, too. And it was probably incredibly popular for the time as well. But what do we remember decades and centuries later? Only the best.

Want to impress young neighbors or grandkids? Talk to them about THEIR favorite pop-stars on their level. Join them in their video games and better yet, learn how to BEAT THEM at their own games! Know what shows they are watching or music they are listening to. Understand their level of wisdom, join them in it. Then, when they let you, share little bits of your wisdom with them. They can't take much of it at a time, so go sparingly. Wisdom is like Dietary Fiber. It's necessary for a healthy life, but too much of it at once just sounds like a lot of gas. It takes some sweetener to make kids want to take medicine. A spoonful of sugar helps the medicine go down.

Way back, near the beginning of this book, I mentioned Tony Robbins' list of human needs. The 6th one I withheld from you at the time is this: Contribution. We want to create something for others to enjoy.

It's easy to think that the world is getting worse, or that society is losing its morals and will fall apart soon. What's more likely the case is that the world has ALWAYS been a hard place to live, but as we get older and more aware of what is going on around us, we see it as new because it is new to us, especially as we become more connected through the internet and modern technology. I saw a Meme long ago, it was a picture of a fake book titled, "The World Was Always Awful: A Guide to World History for People Who Romanticize the Past." If you don't believe me that the world is getting better, I STRONGLY encourage you to go find Hans Rosling's book, "Factfulness."

It begs the question, why do we think the world is so awful, when it's always been this way, or, worse? Why are our standards for the world so much higher than reality dictates? Why are our expectations so different from reality? Perhaps our parents did a great job protecting us from the traumas of the world when we were little. Perhaps we've sheltered ourselves into our own bubbles, whether it's the bubble of our urban city-life, or the bubble of our church, or community, or social circle. Perhaps, all we "Know" about "the other half" is the skewed caricature we see in TV, Movies, or the News.

The fear that the world is getting worse causes us to lose hope in our fellow humans and humanity in general. We become mistrustful. We hang onto our bad experiences rather than focus, gratefully, on the huge majority of good interactions that make life even possible. We despair and we stop trying. Instead, we need to find things to Hope for. As you do you will tend to find people who are also trying to do the same. This should boost your hope in humanity, as long as you don't see yourself as better than others for it. If you are "woke" in some aspect, others must be woke in other aspects. As you meet people different from you and respect them and build relationships with them to trust them, you can combine all of it together for an improved world.

Every living creature that is functioning correctly will be working to preserve it's own life and as many other living creatures as it can. Few individuals of any species destroy simply to destroy, intentionally. They exist, and we tend to label those individuals (whether they are humans, dolphins, dogs, or anything) as Evil.

Unfortunately, we may see someone destroy or kill and we don't understand why, so we erroneously label them as evil. Most often, it is because of bad information they were taught by those they trusted, immaturity, or because their mind isn't functioning properly. It really is important to figure out why someone is destructive and harmful if we want to help them, or protect ourselves properly. Most of us are trying to be good, and we will manage to be good most of the time. We need to be careful because it only takes a little bit of damage to undo and destroy what took a great deal of time to create.

For those that believe in God, it's fair to sum up all of God's actions into the purpose of continuing Life over an eternal and supernatural timeline, extending individually far beyond the extent of a mortal life.

At times we need to make hard choices between what is necessary to continue life in the immediate present, and what will help life continue someday in the future. Focusing on immediate needs is often seen as a feminine trait, much like mothers will focus on comforting and soothing a crying baby and ensuring the comforts of a safe and clean home. Setting sights on the future tends to be seen as a masculine trait, like many fathers will encourage their children to push through hard things to make them stronger for the future. Both tendencies can get out of balance and cause damage if taken too far. We need both.

These are general tendencies with huge overlaps between genders. We all grow up best when we can take advantage of and utilize both traits, as a complete set, picking and choosing which trait we need at that time. It's not easy deciding, and we get it wrong a lot. If we keep trying, respecting the opinions and perspectives of each other, the collective decisions of the group will tend to be more right than wrong as long as the needs of the individual are not ignored.

Many of our political differences come down to this conflict: the needs of the immediate moment against the needs of the future. Or the needs of the individual against the needs of the society. Many of our family problems, too. It's easy to see a problem and come up with some immediate way to solve the problem, only to find out later, that hundreds of other people had already thought of that idea, explored it and tried it, and realized it doesn't actually work. Don't become Co-Dependent on Society or your family to fix itself for you to find happiness.

Take responsibility for your actions. Man-up, or Woman-up. Embrace who you are at this moment, and then take the steps to Grow-up. For a long time, I had this theory that "People don't actually grow up or mature past high-school, they just get better at hiding their immaturity around strangers and kids." I still feel there's a lot of truth to this, but I've learned and observed that many people do manage to grow up and mature past their High School levels.

Growing up, Adult-ing, and being responsible is HARD! Relationships are HARD! Learning to impress people with your career skills or personal skills, and then convincing strangers to trust you to let you help them is TOUGH! INCREDIBLY TOUGH! It's hard enough without any of our addictions or bad habits complicating things. Get outside yourself, find out what you need to to help people be able to trust you, make the hard decisions to let go of parts of yourself that are hurting you and others, and find ways to become better. Some people call this a Purification or Sanctification process. Repenting. Becoming closer to or more like God. More like our Ideal.

If looking for pleasure and stimulation for a brief moment gets you through a rough time, you may need to go with that. If you need to consume more than you create for a period of time, you might just need the break. Just keep it in the back of your mind, that eventually —hopefully sooner rather than later—you'll need to get back on the train and keep working to help yourself and those around you, responsibly. Remember the whole time, that the longer you wait, or the more destructive the break, the harder you're going to have to work to repair the damage, and there may be permanent and lasting damage that won't ever go away. You will need to find a way to help others, WHILE, you help yourself.

If you get a chance, go find the TED talk by Jill Bolte Taylor as she, a brain scientist, describes what she experienced and observed in herself while she suffered a stroke. There are two focuses of our mind. It can focus on internal things or it can focus on external things. As Jill suffered from the stroke, she could feel those two aspects cutting in and out as blood flowed or was cut off from different regions of her brain. One moment she would feel like she was "one with the universe" and suddenly come crashing back to individuality. It's very entertaining and enlightening to hear her describe it.

Her story portrays the duality of each of us: our connection with everyone else and our need to preserve ourselves. We cannot exist long without one or the other. Helping each other while we help ourselves, using some fixed reference point, such as continuing life, or happiness, or spirituality, or ethics, is crucial and is an ingrained part of who we are as Homo Sapiens: As Humans. It is built into our DNA. Every culture on this planet has some form of the Golden Rule: Do unto others as you would have them do unto you. Or better yet: Do unto others as THEY would have done unto them.

Moralists, ethicists, and philosophers can spend all day arguing and discussing circles around each other about how to best help others and find happiness. One goal of spirituality and religion is to give you an external perspective on yourself and the world, to see yourself from a third-person point of view. How does God see you? How does society see you? How does Nature see you? To look at yourself like an outsider sees you. That kind of information is invaluable and worth more than just about any other kind of information we can get. "To see as we are seen."

Having an external point of view, a fixed and immovable perspective, is like having a flagpole hold up a flag in the wind. No matter the direction, the flutter, or the speed of the wind, the flag will stay put and all that strain and pull on the flag only displays the image better. When the flag lets go, or fails, it and its message are lost from the skies, only to fall, crumpled on the ground. Religion and spirituality are that fixed point, that flag pole, for many people. Others call it Morality or Ethics. To most people, there's some crucial differences. Whatever you call it, you and I all need some fixed point from which we can examine our own life from others' point of view. And we need to understand each others' fixed points as we help each other strive for improvement in our own ways and help them with theirs.

Sharing what you have learned is good for you and good for others. It keeps you in check and serves to surround yourself with others who are improving their lives, too. They, in turn, can surround themselves with others who are improving their lives. Biologically, as a species, this is how humanity has thrived: Respecting individuality, while helping and serving each other.

It's a bit like being in the habit of inviting company to your house. You are more likely to keep your home clean and bright and inviting if strangers, friends, or family, come over regularly. This gives them a safe and fun place to spend their time, and when they are not around, it gives you a warm and comfortable place to spend your own time.

Our bad habits and addictions have the tendency to isolate us from others, eventually. They cripple our ability to express ourselves in appropriate ways, or, they hurt our ability to think clearly and logically. They encourage behavior that eventually drives people out of our lives. When we can remove those bad habits, destructive thoughts, and addictions, we naturally develop the tendency to invite people into our lives to stay, not out of coercement, or guilt, or dependence, or selfishness, but out of love, and safety, and happiness. Rather than a revolving door of relationships we find stability and security in relationships and through that security we find the greatest satisfaction and collective improvement of society.

When you find a part of your life that is suffering, use these same ideas to make the change you need to. Whether it's too much time watching sports at the neglect of your family, or being a bit risky in your driving, find some way to get your head out of the sand, and address your problems, even (and especially!) if you like those problems. Get outside of yourself, give that part of yourself up, and learn that as you do, you'll discover more about yourself and your individual strengths that make you proud to be who you are and eager to help others be their best, too.

As I wrap up this book, think about whether all of this information has helped you. If it has, an easy way to help others, is to share this book with them. Encourage them to read it. Teach them what you've learned from it. Get comfortable letting others know you care about them and that you've found something good in your life and it's made your life better.

Recap: Live and help others live.

30

How to Get Along With People You Can't Stand

Let's be honest. This world would be a whole lot better without all those jerks filling it up, wouldn't it? We all could be decent honest people if it weren't for those idiots causing so many problems. In fact, we all ARE decent, honest people. It's all those OTHER people who are indecent and dishonest! I mean, why do we even have to put up with them, anyway? We need to just stifle them all, or kick them out, or, or, or, maybe we just need to move to Canada every time the Other party wins the election!

All kidding aside, it's really hard to get along with people who are wron- I mean, people who think differently. Bruce Christopher, a well-known psychologist, counselor, and speaker promotes his idea of "The Six Difficult Personality Types." He gives his audiences a really good primer for how to deal with all those difficult people life throws at you. From the temper-tantrum-throwing "Exploders," to the shy, timid, indecisive "Clams." From the friendly and entertaining, yet non-committal "Charmers" to those "Snipers" who love to zing you with jabs and insults across the dinner table or the conference room. He gives amazing insight into how to deal with and work with those people who are hard to work with. He also encourages everyone figure out which type, or types, of difficult people you yourself are.

The most important part of dealing with any difficult person is to remember that anyone becomes a "difficult" person when they don't feel safe; when they are insecure or threatened in some way. Usually they are not even aware they are feeling insecure. Their "difficult" personalities are their defense for feeling insecure. A secure and confident person doesn't have a problem making a difficult decision, or keeping a commitment, or keeping control of their temper and their insults. Every bully on this planet feels unsafe in some major way. Even much of the Nazi rise to power and aggression in Germany leading up to World War II was the result of the insecurity of the German people from the aftermath of World War I. Much of Japan's own expansion in the 1940's was to pursue fertile lands for growing the food that they were feeling was in too short supply for themselves. That leads us back to the Maslow Hierarchy of needs. While the order of Maslow's pyramid isn't true for quite everyone, those needs are universal and everyone feels insecure without them. And we all become difficult and irrational when we feel unsafe or held back.

Even the most rational, scientifically sound mind, the most analytical or unfeeling person, the most dignified king or queen will act like an animal when something surprises and scares them. Like the famous boxer, Mike Tyson, is known for saying, "Everybody has a plan until they get punched in the mouth."

When you encounter someone you cannot figure out how to work with, do some research. Dig deeper into that person's life (if it's appropriate) or dig into psychology and Emotional Intelligence to figure out how to help them. Or if that's not appropriate, at least research ideas of how to work with or live with them without resorting to silence or violence. Figure out that third door between withdrawing and attacking that builds relationships instead of destroying them. It will be a bit different for each situation, though there are some commonalities. I encourage you to go search out the "Crucial Conversations" books I mentioned earlier. Or Bruce Christopher's lectures. And learn to stand your own ground, AS LONG AS you are not the one under the influence of some addictive behavior or substance. If you know that you might still be under the influence of a hangover, or medication, or loss of sleep, or "time of the month," hold your peace just a bit longer until you KNOW that you are as clear-headed as can be.

Only when your own mind is clear and you're not under the influence of some hormonal surge, some drug or behavior, or dealing with some emotional trauma yourself can you think clearly to help others. Only then can you know for sure that the problem isn't with you. Only when you've removed the "beam" from your own eye can you see clearly how to remove the "splinter" or "mote" from another person's.

If you're in a difficult relationship and you recognize that both of you are being difficult, find some way to get both of you the help you need, but get your own help first and encourage the other person to come along with you. Keeping relationships healthy will be the most creative endeavor you can undertake in this life. If you don't want to "die with the music still in you" —as Oliver Wendall Holmes is credited with saying— use all your ability to build relationships in your own unique way, but recognize you'll have to build the relationship from where the other people in your life currently stand as well.

When a new connection is made between nerves and neurons within our brain, we laugh and feel good. When those connections are lost or destroyed, we often feel sad or depressed. The same is true with humanity and the world. As we form connections with others, we feel fulfilled. When we lose them, we feel empty. All of this book is an attempt to help you and everyone around you form new connections, with yourself and with the rest of the world.

In the end, we will all need to learn to trust people that we naturally may not want to trust. Our instincts, fears, drives, and genes may push us to be wary, hesitant, and scared of those that we see as different from us, or that have a natural authority over us. We will be mistrustful of people who have hurt us, or who belong to a group that has hurt us. No matter who you are, you will be hurt by those who are supposed to love you. We are all too clumsy and imperfect to never hurt those closest to us.

In fact, it is those closest to us that we will likely hurt the most. Like rocks getting tumbled down a fast-moving river, or like a fleet of ships in a storm, we will bump and jostle, and damage each other just by being close to each other. And in this jostling, eventually learn to fit in and settle into useful places. We start to fit together like stones in a fortress wall. Where we before bashed against another person, and then held on tightly, we now fit together like puzzle. Sometimes, we will bash too hard and may need to distance ourselves to keep from being completely broken or destroyed, but for those relationships that can survive and learn to repair the excessive damage, we become stronger together.

We are all in this together and with as widespread as humanity has been for who-knows-how-long there's no escaping from each other. Even our wilderness areas are protected by rules and regulations to keep them that way for everyone's enjoyment and use. We try to hide from each other in our homes and bedrooms, but we end up streaming and scrolling through endless entertainment, or media, created by others, for our benefit, on technology created by thousands and millions of people working together, towards common goals. Led by those who are good at inspiring and driving people to overcome their differences and push forward united in a purpose, in a country that allows us to have some trust in the system and incredible freedoms and luxuries never before enjoyed by humanity. Even the wealthiest kings of medieval Europe, or ancient China, didn't have so many of the amazing privileges or as advanced comforts and technologies as most of you reading this will have.

However bad you have it, please take a moment to look around, and find things to enjoy. Things that have made the memories you enjoyed. Things that have given you comfort, safety, and choices. Imagine and think of all the countless people who worked, sacrificed, and served each other to get you all that you have, or have had.

To get along with people you can't stand, find some unifying goal you share with that person. If it's a love for your community or country, start with that. If it's a goal to keep a marriage and family together, go with that. Build on that. Those are great foundations for relationships to start on. No one will trust anyone else if they don't feel safe and secure. Anger, yelling, name-calling, insulting, these do a great job of getting people to retreat into their shells. Think of a scared snail, turtle, or rolly-polly curling up into their shells when scared. Only when someone feels safe will they come out and cooperate.

Next, address the points of conflict, ONE specific thing at a time. Don't drag the past in, deal only with the present. Let go of the baggage. Keep your discussion factual. Don't point fingers and blame the other person or demand they change. Address the source of the problem. "When you talk to me like that, it makes me feel like you don't love me."Listen, and work to understand the other person. Put aside your worries and concerns and do your best to understand the other person, even if they're wrong and you know it, understanding them will allow you to connect with them and help them move forward.

Sometimes, people have been hurt so many times they will no longer tell you what's wrong. John Gottman calls this "Stonewalling." When a person shuts down. Often times, marriages don't fall apart when the couple is still fighting, it's when they quit fighting that the marriage has finally gone sour. People who are arguing with each other are still caring enough about each other and the common goal to put in the effort to come up with a resolution.

Getting someone to open up when they've shut down takes patience, care, and some sensitive skills. First, ask them what's wrong. If that doesn't work or they don't seem to be telling the truth about what's bothering them, "Nothing's wrong, I'm fine," then point out to them and hold up a mirror of their own appearance and behavior. "You don't look fine, everything about you tells me that you're upset about something."

If that doesn't work, try to paraphrase their previous complaints. Repeat their own words back to them. Make sure you heard them right, then rephrase it into your own words to make sure you understand them correctly. Let them fix your understanding until you are both on the same page. This can take some questioning and work just to make sure you are even talking about the same thing.

If you are still stumped, the snail is still hiding, and you can't seem to get to the true source of contention, prime the pump. Guess at what might be wrong. Make your best assumption and ask if that could be the problem.

If you are the one who feels unsafe and attacked, and you see that someone is trying to figure you out, understand that they are showing care in the best, clumsy way they know how. Help them as best as you can.

Do your best to avoid fighting or running away. The natural instincts are Fighting and Flighting. Silence and violence. Look for the third option, somewhere in between where connections can get re-established and relationships made. This is the only way that humanity has progressed as far as we have. We see a problem and we work together with others to solve it to help as many people involved as possible. We cannot please everyone, though we should try to please the most people we can.

The core to all this: Be Nice

31

Distraction, the Last-Ditch White-Knuckle Effort

Sometimes, even when we know how to handle our instincts, our drives, and our reflexes we still get blindsided. Sometimes, we get overwhelmed with triggers, or pent-up hormones, and we act out despite all of our best intentions, solid relationships, support systems, back-up plans, and everything else. Sometimes this world is just too much for us and we need a way to shut it out, drown it out, or escape it and our responsibilities to it for a time. This is when we get into trouble if we don't do it carefully.

Sometimes we just need a distraction.

Sometimes, as a last-ditch effort, we need an emergency line we can reach out to. A close friend, a quick call. We need to just endure it sometimes. Sometimes we call it White-Knuckling because you're just holding onto sanity as tightly as you can.

When I can catch my mind and thoughts trying to keep going places it shouldn't go, sometimes I can get on top of it by using my dominant sense, my eyes, and calling out, either out loud or in my head, what I'm seeing. "Blue car." "Six orange cones." "Snowy mountain." "Save-all Storage."

"Two people on the sidewalk." I keep this up as long as I can until more distracting things demand my attention. Some people find that reciting poems works for them, or singing songs. Some people go exercise real hard. Focus on how fast you are breathing, how hard your heart is beating, or how tense you are becoming. Slow down your breathing, let your heart calm itself. Mentally search your body from head to toe and find any muscles that are tense that just don't need to be. Focus on them to help them let go and relax. The idea is to use your senses and behaviors to occupy your mind, rather than letting your mind control your behaviors.

Sometimes, Life and Biology still just gets the best of us and we mess up. Even after amazing progress, experience, and learning.

We are all grossly imperfect. Anyone who can't understand that, who doesn't feel their own guilt from their own struggles and failures, doesn't have enough self-awareness to realize the damage they themselves are causing. Nothing fixes a judgmental person like finally getting busted for their own destructive habits that they've managed to hide, sugar-coat, rationalize, or just plain avoid consequences for, and getting it shoved back in their face. "This is YOUR TURD! You made this mess, yes YOU!"

A person stuck in their instincts and reflexes will still ignore or explain away their faults as being other people's problems or as non-existent, or not-serious enough to demand attention. Or they will lash out in anger at the person exposing them to their faults.

A wise person, a human being, will acknowledge the truth, fess-up, and then work to fix it as best as they can. In doing so they become a wiser, more mature and grown-up person. The incredibly successful play "The Fantasticks" illustrates this amazing life lesson so well, all with funny, silly, and deeply thoughtful songs and dialogue. This theme is also the driving core of the musical, "Into the Woods." We cannot grow up without messing up and learning from it. We just can't.

A baby cannot learn to walk without falling down, a lot. A child can't learn to ride a bike without crashing several times. Most horse riders will tell you that you are going to get knocked off more than a few times, even after you've gotten good at it. If you haven't fallen so hard it makes you wonder whether you should even get back up, just wait, it'll happen to you, too.

For those of us who fail after trying so hard, it can be crushing and incredibly discouraging. Soul-crushing, even. Our striving for perfection and crashing and eating dirt hard can be hard to deal with. The important thing is that we pick ourselves up, dust off our hands and our pants, and get back on. We keep trying. We keep working and learning. We keep helping others. And we try again.

Children will not learn to respect their parents if their parents give in to their demands, a horse will not respect its rider if it can buck them off and get away with it, and our own body will never get under our own control if we quit trying.

I thought I was nearly finished with this book, when I had a rough night of my own, and thought about it all morning, before writing this and another chapter. There's still more I need to learn, like you. We are all in this together. That's why I wrote this book, to help myself and as many of you as possible.

Let me sum up: Hang on.

32

Conclusion

A lot is said about happiness, inner peace, enlightenment, heaven, love, nirvana, ecstasy, and joy. I said earlier that the purpose of life is to Continue Life. It is no surprise that the greatest physical satisfaction we can find comes from our sexuality, or some imitation of it. It is also no surprise that to many people, the greatest emotional joy they find in life is found in raising their children or helping others who are seriously struggling, as difficult and painful as those processes are. For most of the world's major religions, perpetuating Life into the eternities of the universe is the primary goal. They spend most of their time trying to teach their adherents to live sustainably good lives while setting up ideals to aim for. These ideals usually also happen to allow life to thrive on this planet. Or, they do this as best as imperfect people can.

Just about every endeavor and action, and ounce of energy we spend is our attempt at achieving the continuance of life, or working toward that. Understand that this is a universal goal of almost every person you will ever meet. They are working towards that Goal the best that they know how. It's usually only when we've been repeatedly unsuccessful at that goal that we quit and pursue simple selfish pleasures. When despair sets in and we lose that last bit of hope in anything good we begin to hate ourselves and the rest of humanity. Self-loathing, self-harming, and the dislike of humanity in general is a sign of a mind that having trouble finding things to hope for. If you can recognize this in yourself, use it as a sign that you need help.

We all should want happiness or fulfillment. Nearly all of us want at least a bit of it while we're still alive. Some of us are willing to work towards receiving that fulfillment in a next life.

Some of us were raised in a home that taught us that we were nothing, that we deserved to be beaten down and abused, that we are only worth the money, time, or energy that someone else is willing to spend on us, that the only joy in life is to be found in any cutthroat way you can get it, or something that stimulates as many of our senses and nerves as possible. Some of us weren't even raised in a home, as sad as that is.

Some of our homes were more like palaces on the outside, but emotionally like empty tombs on the inside. When we are raised in that kind of environment, our brain, our body, and our soul still yearns and needs to feel all those positive things I mentioned in this book. It will find anything it can that imitates, simulates, copies, or creates those feelings and hormones and it latches onto them out of desperation, whether it's healthy or not. When we haven't been taught to watch out for those things, or worse, if it was those raising us who taught us to pursue them, we are starting life at a disadvantage. In a trap. Like being born in quicksand.

Many people on this earth will not escape that quicksand. There have been and will be millions who will be born, raised, live, and die in a mire of selfish indulgence, or born suffering the effects of others' selfish indulgences. There are millions who will only interact with other people in order to use them to satisfy a physical urge or financial gain and nothing more. That is the way of this earth and biology and life. Morality, Religion, and Spirituality exist to combat this natural decay.

They don't always succeed, and perhaps even rarely succeed, but that they exist and that there are those out there determined to keep trying, says something about us as a species. We have the ability to overcome our natural instincts, our base desires, our primitive animal selves. We can be more than just a biological creature. We can get better. We can improve. We can spend our energy helping others without thought of reward. Some cynics will claim that the promise of a Heavenly reward is a selfish one, but since most of us never get a taste of Heaven, I'll argue, it's not much of an enticement, at least not for very long. Being unselfish and helpful for an entire life is really hard.

What we truly can find is that the serving and loving of others, IS heaven itself, and we begin to hope and look forward to a Heaven, or Paradise, or Enlightenment where all we can do is love and serve others and help them love and serve us in return.

The hope for that kind of Heaven or Utopia is what drives us onward, to keep trying, despite the incredible numbers and statistics and history that try to show us we are failing. As we work against such an opposing storm, as we fight against the chaos and entropy that is integral to the Universe, we see that beneath the raging storm of failure, suffering, and injustice, are the safe and inviting homes, huts, and fires burning in cultures and peoples all across this planet.

What we see as injustices are often the fuel for those fires. What we see as suffering is the fertilizer for the growth of magnificence. What looks like failure on the outside is the guiding hand of our universe determined to point us in another upward direction or away from danger, sometimes using us as the example of what NOT to do. Rest assured, that most people are good, most of the time. If this were not the case, humanity would never have made it off the savannas of ancient Africa. Somewhere between the purely instinctive hunting-gathering phase of our species and the farming and domesticating of the Earth we gained something no other animal had ever had before. We have something that inspired us to be more than animals, and it has been fueling our progress ever since.

We cannot see these things if we are lost in the storm ourselves, or if our focus is only on the destruction the storm is causing. Some of us will deal with droughts and fires, some with floods and hurricanes. Some of us will deal with cold, some with heat. Some of us will fight and struggle against a wilderness to scrap out a meager living. Some of us will be born in a soft and non-challenging paradise surrounded by comforts and luxuries and fail to develop empathy and understanding. Some of us will feel the crushing blows dealt to us by those who've lived under those trials, the reflexive lashing-out of those born in harsh circumstances who weren't able to learn any better.

There will be breaks in the storm, there will be times of calm. There will be peaceful moments. Unfortunately, as long as we live in a universe where entropy, chaos, and disorder are the final results, we will be fighting uphill. And the great news is that we have thrived, as humans, against all of it for thousands and millions of years. And for the last 5 thousand years of our history, this progress has accelerated exponentially and shows little sign of stopping. Should our civilization ever collapse, we have the records, the information, the knowledge that will stick around to help us rebuild it, and make it even better. I don't think Humanity will ever cease working towards Heaven. It is in our DNA, our environment, and our collective decision-making ability. Individually, we will struggle. Globally, we will struggle. That has to be the case. There is so much to learn and each new generation has to learn it from scratch as well. Collectively, we're doing pretty good.

There's a fable that tells how each of us has two wolves inside of us, fighting for prominence and domination. One is full of hope and light, the other full of despair and darkness. For most of us, these two wolves are evenly matched. Sometimes in our life, Hope is in control. Sometimes Despair is. The moral of the fable is that the wolf that's winning is the wolf that is being fed. The difference between them is where these two wolves draw their power. Are we feeding our Despair, or are we working to feed our Hope? Despair will come at us from all sides. The natural universe will always tend towards decay and chaos. We don't have to go looking for despair, or for food to feed that wolf, it will grow and strengthen purely on its own merits and the forces of nature.

Going against the outward forces of nature, we have something inside of us that is both entirely biological and supernatural that can stir up Hope. We can cultivate it. We can push against the forces of nature, dig deep inside of ourselves, and create something new and organized from the chaos around us. We can build up what nature wants to tear down. Why we do this is going to be different for each of us, individually. Sometimes it will be a desire for financial or material wealth and security. Sometimes it's to preserve our physical, mental, or emotional well-being. Sometimes it's a hope for an afterlife. Sometimes it's just plain habit. Sometimes, it's just what we were taught as kids and is all we know. Our motives will probably change multiple times throughout life. Use whatever motivation keeps you going at that time. As you grow and mature throughout your entire life, the pressures of life will make some of these motivations more relevant than the others. That's okay. Keep creating.

This creative process, this ability to shape and organize matter is what breathes life into us. It is what gets us out of bed in the morning, and what motivates us to spend precious energy and calories as an investment to reap greater rewards later. If you need help getting out of bed in the morning, use whatever reason you can grab onto in the moment and let life keep sorting the rest out once you're up and moving and busy creating. Make conscious decisions when you can.

The battle all of us faces is how to keep Despair from growing stronger, while feeding the underdog, Hope, with whatever scraps of emotional food we can find and cultivate in our lives to keep that wolf in the ring.

Only by working and fighting that uphill battle for Hope can we help each other and save ourselves, and we are not alone. Trust the rest of humanity that millions and billions of others are doing the same in their own way. Former drug addicts can learn to help others in ways that those who've not faced such difficulty can only dream of. Billionaires who need a new purpose or challenge often learn to quietly and secretly give more money away than most of us will ever even see in our entire life. Some of us hash it out online trying our best to share our opinions and perspectives, back and forth. The most effective way to accomplish this building and creating is in careful, stable, ways, but no kindness is ever wasted even if we don't quite get it right.

On very rare occasions we will need to violently tear down an old wall to improve a home, or pull out a cracked foundation. Sometimes we will need to revolt against a system in defense of our lives, our families, our homes, and our freedoms. The great most times, though, we are just redecorating, refining, and adjusting. Peace and cooperation, dialogue, win-win arrangements, and building each other up, have the greatest effect, almost every time. Sometimes we have no choice but to just wait and Hope for others to resolve a problem for us, save us from the predicament we are in, or for them to solve their own problems when they've refused our help.

Make life safe for others while you keep it safe for yourself. Respect boundaries while you pull down your walls. Help others do the same. Expand your views, especially when you feel like some problem you are facing is overwhelming, or overwhelmingly bad. Neither are true very often, but when we are stuck in our own world or our own perspectives, they can seem that way. Finding a way to get away from addictive, isolating, or destructive behaviors, and getting a clear head, and control of our frontal cortex —our conscience, our body and spirit— will go a long way to help us empathize with others and gain a clear perspective of reality. The greatest super-power no one talks about is the ability to understand another human being in an amazingly connecting way, and we only get to have that super power when our mind is clear.

The good that Hope can do is incredibly powerful. Sometimes it only takes one good deed, or one vocal opinion to save a person or society from the downward spiral started by those who didn't know better. While one brave leader can break the ground for change, it takes a group or society working together to accomplish it. One person can start the change, and hundreds or millions more need to support it. And not just support it, but be that change, themselves.

When you encounter a dangerous animal in the woods, very often that animal is just as or more scared of you than you are of it. The reason a rattlesnake bites is to protect itself from YOU. The reason bears attack is because they feel threatened by YOU. Bees sting to protect their hive. The same is often true of people. Those that scare us are often just as scared by us. Those that make us uncomfortable are usually uncomfortable around us in return. Be the change you wish to see in the world. Be brave. Be bold. Be Gentle. And be willing to get better. Help others feel safe, and when they feel safe enough, they will almost always want to get better. How safe they need to be will be different for each person. Trust them while you carefully protect yourself. And change yourself to become better.

We all should be planting the seeds of positive change in our own lives. Those tiny seeds, spread out on the wind and multiplied by the tens, hundreds, and thousands, can regrow an entire burnt forest. They can reform a government or religion. They can inspire a nation. They can save millions of lives. Faith, like a tiny mustard seed, can end up moving mountains. You can change the world, one person at a time.

Life is Beautiful. I wish you well as you keep living it.

Notes, References, and Inspirations

Listed below are where where you can find the most of the works I referenced directly, not in any particular order. A great many more sources have contributed to all that I have learned and I wish I could credit all of them.

Joseph Grenny, Kerry Patterson, David Maxfield, Ron McMillan, Al Switzler. "Influencer: The New Science of Leading Change - Second Edition." McGraw-Hill Education, 2013.

Joseph Grenny, Kerry Patterson, Ron McMillan, Al Switzler. "Crucial Conversations Tools for Talking When the Stakes are High, Second Edition." McGraw Hill Education, 2011.

Hans Rosling. "Factfulness: Ten Reasons We're Wrong About the World—and Why Things Are Better Than You Think." Flatiron Books, April 2018

Dr. Seuss. "The Sneetches And Other Stories." Random House, 1961.

Great Big Story. "This Actor's Cartoon Game is Strong." YouTube, July 9, 2018. https://www.youtube.com/watch?v=jpgnMeDjnO8

"Meet the Robinsons" Directed by Stephen Anderson.Walt Disney Animation Studios. 2007

Cal Newport. "So Good They Can't Ignore You." Grand Central Publishing,

2012.

Thomas Stanley and William Danko. "The Millionaire Next Door: The Surprising Secrets of America's Wealthy". Taylor Trade Publishing, 2010.

Tony Robbins. "Money: Master The Game." Simon and Schuster, May 2016.

Ed Catmull, "Creativity, Inc.: Overcoming the Unseen Forces That Stand in the Way of True Inspiration." Random House, 2014

Will Smith, Fault vs Responsibility - https://www.youtube.com/watch?v=USsqkd-E9ag

Smarter Every Day. "Why You Should Put Your Mask on First (My Brain Without Oxygen) #157."
Youtube. https://www.youtube.com/watch?v=kUfF2MTnqAw

Alcoholics Anonymous World Services. "Alcoholics Anonymous," also known as "The Big Book. A. A. World Services, NY, Most recent copyright, 2019

Charles Duhigg. "The Power of Habit: Why We Do What We Do In Life and Business." Random House Trade Paperbacks, 2014

Prager U. "Don't Follow Your Passion." https://www.youtube.com/watch?v=CVEuPmVAb80

Laura Hillenbrand. "Unbroken: A World War II Story of Survival, Resilience, and Redemption." Random House Trade Paperbacks, 2014.

James Christensen. "Fishing"

Seth Stephens-Davidowitz. "Days of Our Digital Lives." New York Times,

Opinion, July 4, 2015.
https://www.nytimes.com/2015/07/05/opinion/sunday/seth-stephens-davidowitz-days-of-our-digital-lives.html?_r=1

Yann Martel. "The Life of Pi". Mariner Books, 2003.
"The Legend of Korra - The Voice in the Night." Directed by Joaquim Dos Santos and Ki Hyun Ryu. Nickelodeon Studios. 2012

Stephanie J Wilson, et al. "Shortened sleep fuels inflammatory responses to marital conflict: Emotion regulation matters." Psychoneuroendicrinology, Vol 79, May 2017, pg 74-83
http://www.psyneuen-journal.com/article/S0306-4530(16)30516-9/fulltext

"It." Directed by Andy Muschietti. Based on the novel by Stephen King. Distributed by Warner Bros. Pictures, 2017.

Marianne Williamson. "A Return To Love: Reflections on the Principles of A Course in Miracles." Harper Collins, 1992. From Chapter 7, Section 3 (Pg. 190-191).

Dale Carnegie. "How to Stop Worrying and Start Living." Gallery Books, 2004

RJ Palacio. "Wonder." Knopf Books for Young Readers. 2012

Leo Tolstoy. "The Complete Works of Count Tolstoy: What Shall We Do Then? on the Moscow Census; Collected Articles. (Volume Seventeen)." Letter to N.N. pg 392. Filiquarian Legacy Publishing, 2012

Malcolm Gladwell. "David and Goliath: Underdogs, Misfits, and the Art of Battling Giants." Back Bay Books, reprint 2015

Malcolm Gladwell. "Outliers: The Story of Success." Back Bay Books, reprint 2011

"A Beautiful Mind" Directed by Ron Howard, starring Russell Crowe. Universal Pictures, 2001

Stephen R Covey. "7 Habits of Highly Effective People: Powerful Lessons in Personal Change." Simon and Schuster; Anniversary Edition 2013.

Shawn Achor. "The Happiness Advantage: How a Positive Brain Fuels Success in Work and Life." Currency, reprint 2018.

Michael Pollan. "Second Nature: A Gardener's Education." Grove Press, Reprint 2007.

Jim Collins. "Good to Great: Why Some Companies Make the Leap and Others Don't." Harper Business, 2001.

Robert Cialdini. "Influence: The Psychology of Persuasion." Harper Business, 2006.

Simon Sinek. "Start with Why: How Great Leaders Inspire Everyone to Take Action." Portfolio, Reprint 2011.

Napoleon Hill. "Think and Grow Rich." Chartwell Books, 2015.

John Gottman. "The Seven Principles for Making Marriage Work: A Practical Guide from the Country's Foremost Relationship Expert." Harmony, 2015.

Sean Covey, Chris McChesney, Jim Huling. "The 4 Disciplines of Execution: Achieving Your Wildly Important Goals." Free Press, 2016.

Joseph Campbell, "The Hero with a Thousand Faces (The Collected Works of

Joseph Campbell).” New World Library, 2008.

“The Shawshank Redemption.” Directed by Frank Darabont and starring Morgan Freeman and Tim Robbins. Castle Rock Entertainment and Columbia Pictures, 1994.

TEDx Talks. “What is Your Polygamy?” by Lance Allred. TEDx Salt Lake City, 2016.
https://www.youtube.com/watch?v=MbXzVrzTXHQ

TED Talks. “Jill Bolte Taylor: My Stroke of Insight.” TED 2008.
https://www.ted.com/talks/jill_bolte_taylor_s_powerful_stroke_of_insight?language=en

Bruce Christopher Lecture Series
https://bcseminars.com/

Jared Diamond. “Guns, Germs, and Steel: The Fates of Human Societies.” W. W. Norton and Company, 2017.

Daniel Kahneman. “Thinking, Fast and Slow.” Farrar, Straus, and Giroux, 2013.

Cal Newport. “Deep Work: Rules For Focused Success in a Distracted World.” Grand Central Publishing, 2016.

Loren Eisely. “The Star Thrower.” Mariner Books, 1979.

Preet Bano Singh, et al. “Smelling Anxiety Chemosignals Impairs Clinical Performance of Dental Students.” Chemical Senses, Volume 43, Issue 6, 5 July 2018, Pages 411–417.
https://academic.oup.com/chemse/advance-article-abstract/doi/10.1093/chemse/bjy028/4996077?redirectedFrom=fulltext

About the Author

Christian Lassen is a husband first and father to five children next. He practices dentistry as a day job, and spends his spare time learning as much as he can about...anything. Biology, relationships, and personal improvement are his favorites.

Made in United States
Orlando, FL
04 November 2026

90205468R00187